PROC TABULATE
BY EXAMPLE

Lauren E. Haworth

Comments or Questions?

The author assumes complete responsibility for the technical accuracy of the content of this book. If you have any questions about the material in this book, please write to the author at this address:

> SAS Institute Inc.
> Books by Users
> Attn: Lauren Haworth
> SAS Campus Drive
> Cary, NC 27513

If you prefer, you can send e-mail to sasbbu@sas.com with "comments for Lauren Haworth" as the subject line, or you can fax the Books by Users program at (919) 677-4444.

The correct bibliographic citation for this manual is as follows: Lauren E. Haworth, *PROC TABULATE by Example*, Cary, NC: SAS Institute Inc., 1999. 392 pp.

PROC TABULATE by Example

Copyright © 1999 by SAS Institute Inc., Cary, NC, USA.

ISBN 1-58025-358-X

Table of Contents

iv

Acknowledgments

This book would not have been possible without a lot of support from a number of people.

At the Center for Health Research, I would like to thank Pierre LaChance and Buzz Rhea for supporting my writing efforts. I would also like to thank Reesa Laws for putting up with my increased stress levels during this writing project.

At SAS Institute, Betty Angell was invaluable in providing constant encouragement, and in handling all of the details that go into publishing a book. Also, Nancy Mitchell's work was greatly appreciated in taking my basic ideas about how the book should look and turning them into reality.

It was a particular challenge to write about a version of software that wasn't even available in beta until the book was nearly completed. I would like to thank developers David Kelley and Paul Kent for their information, insights, and critiques of my early work on Version 7.

Finally, I'd like to thank the SAS user community for all of their suggestions. In particular, the SAS-L contributors gave me a number of ideas about topics to address.

How to Use This Book

Where to Start

This book is aimed at all levels of users, but depending on your level of expertise, some parts may be more helpful than others. Here's a guide to what parts of the book will be the most help for you.

Beginning users: If you are new to SAS software and have little or no experience with PROC TABULATE, you should read "The Basics" in detail. After that, the examples in "Intermediate Topics" will be more understandable, and you can use them as needed. Some of the examples in "Advanced Topics" require more expertise with the DATA step and other SAS procedures and tools. You may wish to save these until you have further experience with SAS.

Intermediate users: If you have already built simple PROC TABULATE tables, you can skim "The Basics" and move on to read "Intermediate Topics" in more detail. Then, you can go on to take advantage of the examples in "Advanced Topics".

Advanced users: If you are already comfortable with PROC TABULATE table production, you will want to skim "Intermediate Topics", and then go on to read "Advanced Topics" in more detail.

All users: No matter what your level of expertise, you should take a look at "Common Errors" whenever you run into problems. The "Special Topics" chapters on options and FORMCHAR settings are also an important reference. Depending on the version of SAS software that you are using, you may find the chapters on Version 7 enhancements and SAS/ASSIST useful as well. Finally, the "References" lists other books and articles you may find helpful.

How to Use the Examples

You don't have to read this book from cover to cover to take advantage of the PROC TABULATE techniques it contains. The book is set up as a series of examples, so you can zoom in on the information that best meets your needs.

The description of each example and the SAS code are always on the left side of the page, and the example output is always on the right side of the page. Everything you need to understand each section is right in front of you, so you don't have to turn any pages. When you need help creating a PROC TABULATE table, follow these steps:

1. Find the chapter that seems to cover the general type of table you're interested in. For example, if you need help with percentages, try Chapter 7, "Handling Percentages," or Chapter 8, "Handling Percentages with Complex Denominators."

2. Flip through the chapter looking at the example output until you find a table that looks roughly like the table you're trying to build.

3. After reading the explanation, use the example code to start building your own table. Figure out which variables you want to use for your rows, and substitute them for the row variables in the example code. Then figure out which variables you want to use for your columns, and substitute them for the column variables in the example code. Finally, figure out which statistics you want to use, and substitute them for the statistics in the example code.

4. Do a test run of your code to see if you've got the approximate table that you were looking for. If you run into problems, reread the explanation. If you're still stuck, refer to the chapters in the "Common Errors" section.

5. Once the basic structure is correct, you can work on cleaning up and refining your table. Chapters 9–13 cover topics related to revising the appearance of your table.

Tips, Warnings, and New Features

Throughout the text you will see tips and warnings related to some of the examples. Look for distinctive icons placed in boxes separate from the rest of the text. Be sure to read any warnings associated with the examples before you try the code in your own programs.

 This is the symbol used to identify a warning. Warnings generally contain additional information about potential risks associated with the technique used in the example.

 This is the symbol used to identify a tip. Tips generally contain additional information about the example, or explain an alternate technique that can be used to create the example table.

 This symbol refers to enhancements available in Version 7 of SAS. Sometimes, there may be an easier way to build a table if you have Version 7, and it will be mentioned using this symbol. Detailed information about Version 7 is included in Chapter 24, "Version 7 Enhancements."

2

CHAPTER 1

Why Use PROC TABULATE?

SAS software provides hundreds of ways for you to analyze your data. You can use the DATA step to slice and dice your data, and there are dozens of procedures that will process your data and produce all kinds of statistics. But odds are that no matter how you organize and analyze your data, you'll end up producing a report in the form of a table.

Whether your reports are for an article in a scientific journal, a market analysis for a client, or an internal report for your boss, what you need is a concise table with all of the pertinent results in one place.

This is why every SAS user needs to know how to use PROC TABULATE. While the TABULATE procedure doesn't do anything that you can't do with other procedures, the payoff is in the output. PROC TABULATE computes a variety of statistics, and it neatly packages the results in a single table.

Instead of running two or three procedures and then having to either turn in your results as a big stack of output or retype the results into a table, you can use PROC TABULATE to create a single piece of output that's ready for delivery.

To illustrate the similarities and differences between PROC TABULATE and other SAS procedures, the following example takes the same analysis and produces the results, first using PROC MEANS and then using PROC TABULATE.

Example without Using PROC TABULATE

Here's the situation: your boss wants to know the average age, income, and education of your customers, overall and by gender. "No problem," you say, "SAS software can do that easily."

To get the overall means, you use the following code:

```
PROC MEANS;
    VAR AGE INCOME EDUC;
RUN;
```

This produces Output 1.1. The column under the heading MEAN has your desired results. But wait a minute, didn't your boss also want the results broken down by gender? Now you have to add another procedure to your program. To get the breakdown by gender, you repeat the same PROC MEANS, adding a BY statement:

```
PROC MEANS;
    BY GENDER;
    VAR AGE INCOME EDUC;
RUN;
```

This produces Output 1.2. Your additional results are shown in two columns, one under GENDER=Female and one under GENDER=Male. So now you've got everything your boss wants. The only problem is that it is scattered across three different places in two separate pieces of output.

You can retype the data into a table in your word processor, but that takes time and may introduce errors. Or, you can give your boss the pages of output with the relevant results circled — not a very professional solution. Or, you can turn to a different procedure that's more suited to the task: PROC TABULATE.

Example Using PROC TABULATE

To produce exactly what your boss wants, use the following code:

```
PROC TABULATE;
    CLASS GENDER;
    VAR AGE INCOME EDUC;
    TABLE (AGE INCOME EDUC)*MEAN, GENDER ALL;
RUN;
```

This generates Output 1.3. It has all of the numbers your boss wants in a single table and is ready for delivery.

Okay, so if PROC TABULATE is so great, why doesn't everybody use it? The answer is that many users find PROC TABULATE hard to learn. The syntax is not as intuitive as some other SAS procedures. The goal of this book is to debunk the myth that PROC TABULATE is impossible to learn. The following chapters walk you through the process of building PROC TABULATE tables step by step, with plenty of examples. Then when you're ready to build your own tables, all you need to do is copy one of the examples and modify it to suit your needs.

Output 1.1

Variable	Label	N	Mean	Std Dev	Minimum	Maximum
AGE	Age	6639	48.614	16.598	25.000	90.000
INCOME	Income	6639	25065.797	23850.488	0.000	263253.000
EDUC	Education	6639	13.040	2.953	4.000	19.000

Output 1.2

GENDER=Female

Variable	Label	N	Mean	Std Dev	Minimum	Maximum
AGE	Age	3559	49.528	17.158	25.000	90.000
INCOME	Income	3559	17780.087	17070.596	0.000	263253.000
EDUC	Education	3559	12.932	2.899	4.000	19.000

GENDER=Male

Variable	Label	N	Mean	Std Dev	Minimum	Maximum
AGE	Age	3080	47.558	15.864	25.000	90.000
INCOME	Income	3080	33484.577	27520.481	0.000	251998.000
EDUC	Education	3080	13.165	3.011	4.000	19.000

Output 1.3

		GENDER		
		Female	Male	ALL
Age	MEAN	49.53	47.56	48.61
Income	MEAN	17780.09	33484.58	25065.80
Education	MEAN	12.93	13.17	13.04

PROC TABULATE Syntax

Here are the basic elements of PROC TABULATE. Don't worry about understanding the syntax right now. Use this chapter as a reference when you need it.

Basic Syntax

```
PROC TABULATE <option-list>;
  CLASS <class-variable-list>;
  VAR <analysis-variable-list>;
  TABLE <<page-expression,>
     row-expression,>
     column-expression
     </ table-option-list>;
  BY <NOTSORTED> <DESCENDING> variable-1
     < ... <DESCENDING> variable-n>;
  FORMAT variable-list-1 format-1
     <... variable-list-n format-n>;
  FREQ variable;
  KEYLABEL keyword-1='description-1'
     < ... keyword-n='description-n'>;
  LABEL variable-1='label-1'
     < ... variable-n='label-n'>;
  WEIGHT variable;
```

Much of the above syntax is not required. Each TABULATE procedure requires only a TABLE statement and either a CLASS or VAR statement.

The various options available under PROC TABULATE are listed on the following page. For detailed descriptions of the options, refer to Chapter 21 "PROC TABULATE Options Reference."

PROC TABULATE statement options:

```
DATA=SAS-data-set
DEPTH=number
FORMAT=format-name
FORMCHAR<index-list>='string'
MISSING
NOSEPS
ORDER=order
VARDEF=divisor
```

BY statement options:

```
DESCENDING
NOTSORTED
```

TABLE statement options:

```
BOX=<_PAGE_><'string'><value>
CONDENSE
FUZZ=number
INDENT=number-of-spaces
MISSTEXT='text'
PRINTMISS
ROW=<CONSTANT>|<FLOAT>
RTSPACE=number
```

Available statistics. (These are the basic SAS statistics that are also available with PROC MEANS. See the *SAS Procedures Guide, Version 6, Third Edition* for additional details.) Only the N, NMISS, and PCTN statistics are available for CLASS variables; all other statistics require an analysis variable (listed in a VAR statement):

```
CSS       sum of squares corrected for the mean
CV        percent coefficient of variation
MAX       maximum
MEAN      arithmetic mean
MIN       minimum
N         number of non-missing observations
NMISS     number of missing observations
PCTN      percentage of frequency
PCTSUM    percentage of sum
PRT       two-tailed p-value for Student's t
RANGE     range (MAX-MIN)
STD       standard deviation
STDERR    standard error of the mean
SUM       sum
SUMWGT    sum of weights
USS       uncorrected sum of squares
T         Student's t
VAR       variance
```

PROC TABULATE operators:

, comma: separates dimensions of a table, crosses elements across
   the dimensions
* asterisk: crosses elements within a dimension
  blank space: concatenates elements within a dimension
() parentheses: groups elements so another operator can be applied
   to the whole group
<> brackets: specifies a denominator definition
= equal sign: assigns a label or format modifier to a variable or
   statistic

New Version 7 Syntax

New PROC TABULATE statement options available in Version 7:

```
QMETHOD=method
PCTLDEF=method
CLASSDATA=data-set
PRELOADFMT
EXCLUSIVE
EXCLNPWGT
STYLE=
OUT=
```

New CLASS statement options available in Version 7 (note: you can use multiple CLASS statements):

```
DESCENDING
ASCENDING
PRELOADFMT
EXCLUSIVE
ORDER=order
GROUPINTERNAL
MISSING
STYLE=
```

New VAR statement options available in Version 7 (note: you can use multiple VAR statements):

```
WEIGHT=numvar
STYLE=
```

New TABLE statement options available in Version 7:

```
NOTRAP
STYLE=
```

New statistics available in Version 7:

```
P1, P5, P10, P25, P50, P75, P90, P95, P99
Q1, Q3, QRANGE
MEDIAN
```

New statements available in Version 7:

> CLASSLEV *class-variable-1 <class-variable-2>* STYLE=
>
> KEYWORD *keyword1 <keyword2>* STYLE=

New syntax available in Version 7: styles can now be applied to TABLE statement components, the BOX= option, and the MISSTEXT option:

> TABLE *varname*={LABEL=*'string'* STYLE={*style-attribute1=value*}}
>
> TABLE *varname***statistic-keyword*={LABEL=*'string'*
>   STYLE={*style-attribute1=value*}}
>
> TABLE class-*varname*={LABEL=*'string'*
>   STYLE(CLASSLEV)={*style-attribute1=value*}}
>
> BOX={LABEL=<_PAGE_><*'string'*><*value*>
> STYLE={*style-attribute1=value*}}
>
> MISSTEXT={LABEL=*'string'* STYLE={*style-attribute1=value*}}

CHAPTER 3

Before You Start Writing PROC TABULATE Code

Before getting into the nuts and bolts of writing PROC TABULATE code, let's take a step back and look at the big picture. One of the most common beginner's mistakes in using the TABULATE procedure is to start writing code without a clear plan of action.

This chapter covers the basics of table design. It starts with sketching out a rough draft of the table. The next step is selecting the variables to use and examining the data. It may be necessary to clean some of the data, assign formats, or create categories for continuous variables.

Then comes the task of selecting the appropriate statistics to display in the table. Finally, all of this information can be put together into a detailed mock-up of the table you want to create. Now, you're ready to write PROC TABULATE code.

What Are You Trying to Show?

The first thing that you need to do when designing a table is to figure out what you're trying to show. Let's say your manager just came to you and asked you to build a table that would show whether having a college degree is related to income. This is a very simple example, but it's a good illustration of the basics of table design.

You could generate a printout that shows income for non-college graduates and a printout that shows income for college graduates, but what you really want is a cross-tabulation that shows income and college graduate status in the same table.

Using pencil and paper, instead of a computer, we can quickly draw a sketch of your table to use as a guide for constructing the table. A simple sketch of the proposed table is shown in Figure 3.1.

The next step is to take a look at the data. For this example, and for the remainder of the book, your data is a file containing survey data on education, employment, and income.

Running the CONTENTS procedure on the data set shows us what we have to work with. The output is shown in Output 3.1. The first variable we need for the table is income. Looking at the CONTENTS output, we seem to be in luck, because there's a numeric variable called INCOME.

The next variable we need is one that shows whether the person has a college degree. There's no variable called COLLEGE or DEGREE, but there are two variables that look promising: a variable called EDUC that is labeled Education and a variable called YRSEDUC that is labeled Yrs Educ.

According to the data set documentation, INCOME is the surveyed individual's annual income in dollars. EDUC is a code for the person's highest level of education achieved. YRSEDUC is a recoded version of the EDUC variable, with various levels of education converted into total years of education.

Based on this information, you can select INCOME and EDUC as the variables for your table. However, you're not done yet. There may be other variables that you should take a look at. For example, AGE is one of the variables in the data set. What if this data includes people who aren't old enough to have gone to college? It would be unfair to compare the income of teens who are too young to have graduated from college with that of older people. If there are records for teens in the data set, they should be excluded from the table.

The next thing to look for is basic information about the data set. This is included at the top of the PROC CONTENTS output (and is not shown). Things to look for include the number of observations (6,639) and the sort order (the data set is not sorted).

The last thing to look for is formats. According to the CONTENTS procedure, INCOME and AGE do not have formats, but the EDUC variable has a permanent format assigned. After looking at the documentation, you find out that the format is used to break the values down into five levels of education: <HS, HS GRAD, SOME COLLEGE, COLLEGE GRAD, and SOME GRADUATE.

Figure 3.1

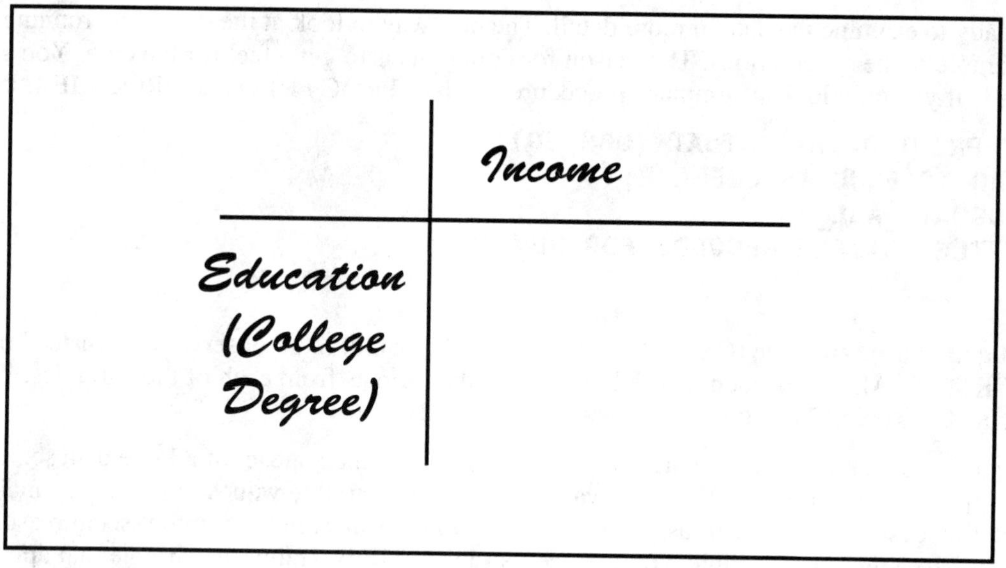

Output 3.1

```
-----Alphabetic List of Variables and Attributes-----

  #   Variable   Type   Len   Pos   Format      Label
-----------------------------------------------------------------
  1   AGE        Num     8     0    8.          Age
  3   EDUC       Num     8    16    EDFT.       Education
 11   EMPLOYED   Num     8    80    EMPLOYED.   Employed
 12   FULLTIME   Num     8    88    FULLTIME.   Employment
 10   GENDER     Num     8    72    GENDER.     Gender
 18   ID         Num     8   136
  2   INCOME     Num     8     8    8.          Income
  6   MARITAL    Num     8    40    MARITAL.    Marital Status
 13   NUMEMP     Num     8    96                No. Employees
  7   OCCUP      Num     8    48    OCCUP.      Occupation
  9   ORIGIN     Num     8    64    ORIGIN.     Origin
  8   RACE       Num     8    56    RACE.       Race
 16   RATE       Num     8   120                Rate
 15   RATIO      Num     8   112                Ratio
  5   SECTOR     Num     8    32    SECTOR.     Sector
 17   UNION      Num     8   128    UNION.      Union
 14   WEIGHT     Num     8   104
  4   YRSEDUC    Num     8    24                Yrs Educ
```

Examining the Data

Now we are ready to examine the data in more detail. The best way to look at the data is by running a PROC PRINT on a sample of the observations. The reason for doing this is to get a feel for the data. You can miss important details if you only look at summary procedures such as PROC FREQ and PROC MEANS.

```
PROC PRINT DATA=TESTDAT (OBS=50);
    VAR ID EDUC INCOME AGE;
    FORMAT _ALL_;
    TITLE 'SAMPLE RECORDS FOR REVIEW';
RUN;
```

To keep the output manageable, the (OBS=50) option limits the output to 50 observations, instead of all 6,639. The FORMAT _ALL_ statement is used to strip off the formats from each of the variables. You want to see the raw data at this point. The output is shown in Output 3.2.

The first thing to do is look for missing data. Though this is only a small subset of a large data set, missing data does not appear to be a problem. Next, we want to look at the variable values to see if they make sense. The documentation describes INCOME as annual income. Do the values in the printout seem reasonable for annual incomes? From this small sample, you can see that INCOME is a numeric variable that appears to be continuous. While one of the values ($34) seems a bit low, the rest seem appropriate for an annual income.

We also want to look at the values for AGE. Do they look like realistic ages, and are these people old enough to have gone to college? At least for this subset of the data, this appears to be the case.

Finally, we need to look at EDUC. We know that this variable has a permanent format, so it's not too surprising that the raw values don't make sense. It will be necessary to use the format when working with this variable.

After looking at the printed observations, the next step is to look at frequency distributions. We can answer a number of questions with a frequency: What do the formatted values of EDUC look like? Is INCOME a continuous variable? Are any of the records for people who are too young to have gone to college?

```
PROC FREQ DATA=TESTDAT;
    TABLES EDUC INCOME AGE;
    TITLE 'FREQUENCIES FOR REVIEW';
RUN;
```

The frequency on EDUC is shown in Output 3.3. This time, the permanent format EDFT is used. With the format attached, EDUC is now broken down into more useful categories: < HS, HS GRAD, SOME COLLEGE, COLLEGE GRAD, and SOME GRADUATE. If you want to see exactly how the format is set up, you could run a PROC FORMAT using the FMTLIB option to get a detailed listing of each of the formats.

The frequency on AGE in Output 3.3 shows that the minimum age is 25, so we don't need to worry about excluding young people. The frequency on INCOME is not shown in this book because it is hundreds of pages long. INCOME is indeed continuous–there are nearly 5,000 unique values of INCOME in the data set.

Output 3.2

```
             SAMPLE RECORDS FOR REVIEW

        OBS     ID     EDUC    INCOME     AGE

         1       1      12     11461      48
         2       2      16     39337      65
         3       4      16     27300      25
         4       6      12     40008      36
         5       7      13     21006      33
         6      10       8     10569      90
         7      11      12     25202      57
         8      12      12     40033      54
         9      13      12        34      52
        10      14      16     10033      26
        11      15      12      9096      70
        12      16      16     21181      69
        13      17      12     18594      27
        14      18      13     32050      42
        15      21      16     28364      47
        16      22      16      7500      46
        17      23      13     10150      40
        18      24      13     58150      41
        19      27      16    100899      40
                                          ...
```

Output 3.3

```
           FREQUENCIES FOR REVIEW

                  Education

                                     Cumulative   Cumulative
        EDUC     Frequency   Percent  Frequency    Percent
    --------------------------------------------------------

    < HS            1001      15.1      1001        15.1
    HS GRAD         2282      34.4      3283        49.5
    SOME COLLEGE    1396      21.0      4679        70.5
    COLLEGE GRAD    1271      19.1      5950        89.6
    SOME GRADUATE    689      10.4      6639       100.0


                     Age

                                     Cumulative   Cumulative
     AGE    Frequency   Percent       Frequency    Percent
    --------------------------------------------------------

      25       138        2.1           138          2.1
      26       120        1.8           258          3.9
      27       146        2.2           404          6.1
      28       157        2.4           561          8.5
      29       170        2.6           731         11.0
      30       185        2.8           916         13.8
      31       170        2.6          1086         16.4
      32       169        2.5          1255         18.9
      33       165        2.5          1420         21.4
    ...
```

Planning the Categories

For most tables, you will want to break down your variables into a manageable number of categories for reporting. For example, a table on age might show categories of 0-19, 20-29, 30-39, and so on. In this table, the variable EDUC needs to be broken down based on college degree status.

For our purposes, we only care if the person has a college degree or not, so the first three categories of EDUC (<HS, HS GRAD, and SOME COLLEGE) can be combined in our table into a single NO COLLEGE DEGREE category. The COLLEGE GRAD category can then be relabeled COLLEGE DEGREE. The only question is what to do with the SOME GRADUATE category? These people have additional education, which makes them different from the people who have an undergraduate degree. It's probably not appropriate to lump these two groups together for this analysis, so these observations should be excluded from the table.

The frequencies in Output 3.4 show EDUC with the new format in place and the people with graduate school education excluded. There are now 5,950 observations, down from the original 6,639.

Our other variable, INCOME, could also be broken down into categories. We could look at how many people have income of less than $50,000, $51,000-$100,000, or greater than $101,000. But then we'd just have a table that shows the number of people with various combinations of income and education. Instead, we can use INCOME to calculate a statistic.

Picking the Statistics

PROC TABULATE gives you a variety of statistics to choose from. We could report the minimum income, the maximum income, or the sum of all incomes, but for this example the mean income seems more appropriate.

To get a rough idea what to expect, we can run a PROC MEANS on the data set to calculate the overall mean income. The results are shown in Output 3.5. The mean, minimum, and maximum values are useful in deciding how to format the table values. Because the the mean is roughly $23,000, the format needs to have a width of at least 7 digits (2 digits for the dollar sign and the comma, and 5 digits for the value).

Drawing a Picture of the Table

Now that we've picked our variables, categories, and statistics, we can turn our rough sketch of the table into a final mock-up. We now know that the row headings will be NO COLLEGE DEGREE and COLLEGE DEGREE. The column heading will still be income, but we can be more specific now and name it MEAN INCOME.

We can also add a title: "Income by College Degree Status" and a footnote: "Note: Data for persons with graduate school education was excluded from analysis." Another thing to include in this drawing is a format for the table values. Because the result is a mean income, the figures should have a dollar sign. Also, we can round off the results to the nearest dollar.

The final mock-up is shown in Figure 3.2. Now you can show this drawing to your manager for approval, and you're ready to start writing PROC TABULATE code.

Output 3.4

Education

EDUC	Frequency	Percent	Cumulative Frequency	Cumulative Percent
NO COLLEGE DEGREE	4679	78.6	4679	78.6
COLLEGE DEGREE	1271	21.4	5950	100.0

Output 3.5

Analysis Variable : Income

N	Mean	Std Dev	Minimum	Maximum
5950	22520.779	20763.443	0.000	263253.000

Figure 3.2

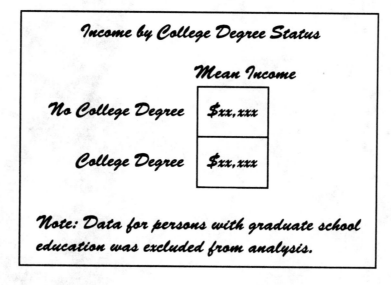

Income by College Degree Status

Mean Income

	Mean Income
No College Degree	*$xx,xxx*
College Degree	*$xx,xxx*

Note: Data for persons with graduate school education was excluded from analysis.

The Basics

20

Creating One-Dimensional Tables

This chapter introduces the basics of producing a table using PROC TABULATE. The simple example table introduces the key concepts behind every PROC TABULATE table.

The Example

This is a basic descriptive table designed to show mean income. It has an overall mean and a breakdown by education level. It also shows the number of observations in the data set.

		INCOME		
		MEAN		
INCOME	INCOME	EDUCATION		
N	MEAN	<HS	HS	College
999	999	999	999	999

To build this example, we will break it down into pieces and build them one at a time. Then the pieces can be put together to build the final table.

Whenever you are building a PROC TABULATE table, it is usually simpler to build it piece by piece. It is easier to see what is and is not working if you don't try to build the whole table at once.

As a first step toward building this table, we will create a simple table that shows income. Next we will add statistic definitions so that the table shows the number of observations and the mean. Then, we will add education level, which will complete the table.

CHAPTER 4

Making the Simplest Table Possible

To understand PROC TABULATE fully, you have to start very simply. The simplest table possible in PROC TABULATE has to have three things: a PROC TABULATE statement, a TABLE statement, and a CLASS or VAR statement. In this example, we use a VAR statement. Later examples show the CLASS statement.

The first part of the procedure is the PROC TABULATE statement. The minimum syntax for this statement is

```
PROC TABULATE;
```

This syntax assumes that the table will be run with the most recently created data set. However, if there is no current data set, you will get the error message shown in the sample log in Output 4.1. A better way to write this statement is to specify the data set to use.

```
PROC TABULATE DATA=TEMP;
```

This code is now more portable. If you want to move it around in your program, you don't have to worry about which data set is the current data set.

The second part of the procedure is the TABLE statement. This tells PROC TABULATE how to build the table. It describes which variables to use and how to arrange the variables. This first table has only one variable, so you don't have to tell PROC TABULATE where to put it. All you have to do is list it in the TABLE statement. When there is only one variable, you get a one-dimensional table, where the single variable is used to create a column.

```
PROC TABULATE DATA=TEMP;
    TABLE INCOME;
RUN;
```

Now our code is almost complete. We've requested a table showing INCOME, but PROC TABULATE needs one more piece of information in order to build the table. You have to specify whether the variable INCOME is intended as an *analysis variable*, which is used to compute statistics, or a *classification variable*, which is used to define row or column categories in the table.

In this case, we want to use income as the analysis variable. We will be using it to compute a statistic. Eventually, we will be computing mean income, but for now we'll just take the default sum statistic. To tell PROC TABULATE that INCOME is an analysis variable, you use a VAR statement to define the variable. The syntax of a VAR statement is simple: you list the variables that will be used for analysis. So now our code looks like the following:

```
PROC TABULATE DATA=TEMP;
    VAR INCOME;
    TABLE INCOME;
RUN;
```

The result is the table shown in Output 4.2. It has a single column, with the header INCOME to identify the variable, and the header SUM to identify the statistic. There is just a single *table cell*, which contains the value for the sum of INCOME for all of the observations in the data set TEMP.

Output 4.1

```
6
7    PROC TABULATE;
ERROR: There is not a default input data set (_LAST_ is _NULL_).
8    RUN;

NOTE: The SAS System stopped processing this step because of errors.
```

Output 4.2

```
 --------------
|   Income     |
|------------- |
|    SUM       |
|------------- |
|166411827.00 |
 --------------
```

Adding a Statistic

The previous table shows what happens if you don't tell PROC TABULATE which statistic to use. If the variable in your table is an analysis variable, meaning that it is listed in a VAR statement, then the statistic you will get by default is the sum. Sometimes a sum is the statistic you want. Most likely, it is not what you want.

And even if you do want the sum of a variable, it's better to specify the statistic explicitly than to let PROC TABULATE select the default. You want your code to be understandable to other programmers, and they might not know that sum is the default statistic. To tell PROC TABULATE what statistic to display, you need to use the appropriate keyword. Chapter 2 has a list of the available statistics and their keywords. If you want to select the sum statistic explicitly, the keyword you use is SUM.

To add a statistic keyword to a PROC TABULATE table, you modify the TABLE statement. You list the keyword right after the variable name. To tell PROC TABULATE that the statistic SUM should be applied to the variable INCOME, you use an asterisk to link the variable name to the statistic keyword. The asterisk is a PROC TABULATE *nesting operator*. Just as you use an asterisk as an operator when you want to multiply 2 by 3 (2*3), you use an asterisk when you want to apply a statistic to a variable. In PROC TABULATE, the term for applying one thing (the statistic) to another thing (the variable) is called *nesting*.

```
PROC TABULATE DATA=TEMP;
   VAR INCOME;
   TABLE INCOME*SUM;
RUN;
```

The output from this PROC TABULATE is shown in Output 4.3. As you can see, it is exactly the same as the output from the previous example. All we have done is make the statistic selection explicit. Now that you know how to specify the statistic, it is a simple matter to change the definition to calculate some other statistic. For example, what if we wanted to know the number of observations for INCOME in our data set? The keyword for this statistic is N. So all you have to do is modify the code as follows:

```
PROC TABULATE DATA=TEMP;
   VAR INCOME;
   TABLE INCOME*N;
RUN;
```

The output with the new statistic is shown in Output 4.4. Note that the variable name at the top of the column heading has remained unchanged. However, the statistic name that is shown in the second line of the heading now says N. And the value shown in the table cell has changed from the sum to the number of observations. This table is the first piece of the larger table that we are building in this chapter.

The next piece of our table is the column showing the mean income. To change the statistic to build this column, all you have to do is pick a new keyword from the list and modify the TABLE statement. This time we pick the keyword MEAN. The output from the following code is shown in Output 4.5:

```
PROC TABULATE DATA=TEMP;
   VAR INCOME;
   TABLE INCOME*MEAN;
RUN;
```

Output 4.3

```
--------------
| Income     |
| ---------- |
| SUM        |
|------------|
|166411827.00|
--------------
```

Output 4.4

```
--------------
| Income     |
|------------|
|     N      |
|------------|
|    6639.00 |
--------------
```

Output 4.5

```
--------------
| Income     |
|------------|
|    MEAN    |
|------------|
|   25065.80 |
--------------
```

Adding a Classification Variable

After seeing the tables we've built so far in this chapter, you're probably asking yourself, "Why use PROC TABULATE? Everything I've seen so far could be done with a PROC MEANS."

One answer to this question is classification variables. By specifying a variable to categorize your data, you can produce a concise table that shows values for various subgroups in your data. For example, wouldn't it be more interesting to look at mean income if it were broken down by the education level of the survey respondent? This is the third piece of the example table for this chapter.

To break down income by education level, we use education as a classification variable. Just as we used a VAR statement to identify our analysis variable, we use a CLASS statement to identify a classification variable. By putting a variable in a CLASS statement, we are telling PROC TABULATE that the variable will be used to identify categories of the data.

In this case, we are going to use EDUC as our classification variable. But before we do so, it's important to make sure that it is a categorical variable. You don't want to use a continuous variable with thousands of values as a CLASS variable because it will create a table with thousands of rows or columns.

So our first step is a quick PROC FREQ to make sure that EDUC is categorical. You can see from Output 4.6 that EDUC is indeed categorical, so we can use it in a CLASS statement in the code below:

```
PROC TABULATE DATA=TEMP;
    CLASS EDUC;
    VAR INCOME;
    TABLE INCOME*MEAN;
RUN;
```

If you look at Output 4.7, you can see that we have left something out. The code above tells PROC TABULATE that EDUC is a class variable, but it doesn't tell PROC TABULATE where to put EDUC in the table. So what we end up with is the same table we had before. If you put a variable in a CLASS or VAR statement, but don't list it in the TABLE statement, PROC TABULATE ignores the variable.

To tell PROC TABULATE what to do with the variable EDUC, you again use the asterisk operator. By adding another asterisk to the end of the TABLE statement and following it with the variable name EDUC, PROC TABULATE knows that the CLASS variable EDUC will be used to categorize the mean values of INCOME.

```
PROC TABULATE DATA=TEMP;
    CLASS EDUC;
    VAR INCOME;
    TABLE INCOME*MEAN*EDUC;
RUN;
```

The resulting output is shown in Output 4.8. Now the column headings have changed. The variable name INCOME and the statistic name MEAN are still there, but under the statistic label there are now three columns. Each column is headed by the variable label Education and the category name < HS, HS, and College. The values shown in the table cells now represent subgroup means.

Output 4.6

Education

EDUC	Frequency	Percent	Cumulative Frequency	Cumulative Percent
<HS	1001	15.1	1001	15.1
HS	2282	34.4	3283	49.5
College	3356	50.5	6639	100.0

Output 4.7

```
-------------
|   Income   |
|----------- |
|   MEAN     |
|----------- |
|   25065.80 |
-------------
```

Output 4.8

```
-----------------------------------------
|                Income                  |
|----------------------------------------|
|                 MEAN                   |
|----------------------------------------|
|               Education                |
|----------------------------------------|
|   <HS     |     HS      |   College    |
|-----------+-------------+------------- |
|   11937.52|   19581.80  |   32710.58   |
-----------------------------------------
```

In general, a CLASS variable that is not used in the TABLE statement is ignored. However, there is a situation where PROC TABULATE does not completely ignore this variable. See Chapter 9, "Handling Missing Data," for an explanation of the relationship between CLASS statements and missing data.

When you are using asterisks to connect analysis variables, class variables, and statistics, the order of the items does not matter.

EDUC*INCOME*MEAN is equivalent to INCOME*MEAN*EDUC. Both will produce the same table values, the only difference is the order of the column headings Education, Income, and MEAN.

Putting It All Together

Each of the tables shown so far is useful, but the power of PROC TABULATE comes from being able to combine several statistics or several variables (or both) in the same table. PROC TABULATE does this by letting you specify a series of "tables" within a single large table. We're going to combine the tables we built in the previous examples into a single table (see the mock-up of this table at the beginning of the chapter).

The first part of our combined table is the number of observations for income. To show this, we'll take a table built as part of the second example.

```
PROC TABULATE DATA=TEMP;
   VAR INCOME;
   TABLE INCOME*N;
RUN;
```

Next, we can add another table from the same example. This table selected the MEAN statistic. To add this table to the first table, combine the code from the two TABLE statements. The code for the two tables is combined by using a space between the two statements. The space (concatenation) *operator* tells PROC TABULATE that you want to add another column to your table.

```
PROC TABULATE DATA=TEMP;
   VAR INCOME;
   TABLE INCOME*N INCOME*MEAN;
RUN;
```

The resulting table is shown in Output 4.9. Note that the additional statistic is shown as an additional column in the table. When SAS is creating a one-dimensional table, additional variables, statistics, and categories are always added as new columns.

The last step in building our combined table is to add the table from the third example in the chapter. The third example introduced the CLASS statement and showed mean income by education. Again we add this information by taking the TABLE statement from the third example and adding it to our current TABLE statement, using a space as the operator between the items. We also have to add the CLASS statement to indicate that the added variable EDUC is being used as a classification.

```
PROC TABULATE DATA=TEMP;
   CLASS EDUC;
   VAR INCOME;
   TABLE INCOME*N INCOME*MEAN INCOME*MEAN*EDUC;
RUN;
```

This code produces the table shown in Output 4.10. Notice how the columns for INCOME classified by EDUC are added to the end of the table.

Output 4.9

Income	Income
N	MEAN
6639.00	25065.80

Output 4.10

Income	Income	Income		
		MEAN		
		Education		
N	MEAN	<HS	HS	College
6639.00	25065.80	11937.52	19581.80	32710.58

Sometimes one-dimensional tables can become very wide. If you're having trouble getting your one-dimensional table to fit on the page, see "Creating a One-Dimensional Table That Runs Vertically" in Chapter 15.

CHAPTER 5

Creating Two-Dimensional Tables

You probably noticed that the table created in the previous chapter was not very elegant in appearance. That's because it took advantage of only one dimension. It had multiple columns, but only one row. It is much more efficient to build tables that have multiple rows and multiple columns. You can fit more information on a page, and the table looks better, too.

The Example

Just as in the previous chapter, this example is designed to explore the relationship between income and education. This time, we are adding two more variables to the table: gender and employment status.

We could build a one-dimensional table to show all of these relationships, but it would have so many columns that it wouldn't fit on the page, which would be confusing to read and interpret. A two-dimensional table is much more appropriate to this task.

		Income					
		Employed PT		Employed FT		ALL	
		N	MEAN	N	MEAN	N	MEAN
<HS							
	Male	999	999	999	999	999	999
	Female	999	999	999	999	999	999
	ALL	999	999	999	999	999	999
...							

(This table has been truncated to fit on the page. There are additional rows for education levels HS and College.)

Turning a One-Dimensional Table into a Two-Dimensional Table

The easiest way to build a two-dimensional table is to build it one dimension at a time. First, we'll build the columns, and then we'll add the rows.

For this first table, we'll keep things simple. This is a table with two columns: one showing the number of observations for INCOME and another showing the mean of INCOME. This table is shown in Output 5.1.

```
PROC TABULATE DATA=TEMP;
   VAR INCOME;
   TABLE INCOME*N INCOME*MEAN;
RUN;
```

To turn this table into a two-dimensional table, we will add another variable to the TABLE statement. In this case, we want to add rows that show the N and MEAN of INCOME for each level of education.

To add another dimension to the table, you use the comma operator. All you do is put a comma between the old part of the TABLE statement and the new part of the TABLE statement.

```
PROC TABULATE DATA=TEMP;
   VAR INCOME;
   CLASS EDUC;
   TABLE INCOME*N INCOME*MEAN, EDUC;
RUN;
```

In the above code, EDUC has been added as a second dimension to the table. But if you look at Output 5.2, you can see that this didn't do what we expected. Instead of adding education levels as *rows* in the new table, it changed the existing columns into rows, and added education as *columns*.

This is because PROC TABULATE expects rows and columns to be listed in a particular order. If a TABLE statement has no commas, then it is assumed that the variables and statistics are to be created as columns. If a TABLE statement has two parts separated by a comma, then it builds a two-dimensional table using the first part of the TABLE statement as the rows and the second part of the TABLE statement as the columns.

So to get a table with income as the columns and education as the rows, you take the code above and reverse the order of the two parts of the TABLE statement. You put EDUC in the first part, and INCOME in the second part after the comma.

```
PROC TABULATE DATA=TEMP;
   VAR INCOME;
   CLASS EDUC;
   TABLE EDUC, INCOME*N INCOME*MEAN;
RUN;
```

This table is shown in Output 5.3. As planned, it has columns for income and rows for education.

Output 5.1

```
----------------------------------
|   Income    |   Income    |
|-------------+-------------|
|     N       |    MEAN     |
|-------------+-------------|
|     6639.00 |    25065.80 |
----------------------------------
```

Output 5.2

```
-----------------------------------------------------------------------
|                          |                Education                 |
|                          |------------------------------------------|
|                          |   <HS     |    HS     |   College   |
|--------------------------+-----------+-----------+-------------|
|Income        |N          |   1001.00 |   2282.00 |    3356.00  |
|--------------+-----------+-----------+-----------+-------------|
|Income        |MEAN       |  11937.52 |  19581.80 |   32710.58  |
-----------------------------------------------------------------------
```

Output 5.3

```
------------------------------------------------------------------
|                              |   Income    |   Income    |
|                              |-------------+-------------|
|                              |     N       |    MEAN     |
|------------------------------+-------------+-------------|
|Education                     |             |             |
|------------------------------|             |             |
|<HS                           |     1001.00 |    11937.52 |
|------------------------------+-------------+-------------|
|HS                            |     2282.00 |    19581.80 |
|------------------------------+-------------+-------------|
|College                       |     3356.00 |    32710.58 |
------------------------------------------------------------------
```

 By the way, there is a limit to which variables you can use in a two-dimensional table. You can't have a cross-tabulation of two *analysis* variables. A two-dimensional table must have at least one *classification* variable. If you think about it, this makes sense. A table of mean income by mean years of education is meaningless, but a table of mean income by categories of education makes perfect sense.

Statistic Definitions for a Two-Dimensional Table

In the previous examples, you may have noticed that the statistic keywords N and MEAN were listed in the column dimension. This is not required. PROC TABULATE allows you to specify statistics in either the row or column dimension (but not both).

For example, the following two-dimensional table displays three statistics for INCOME: N, MEAN, and MAX. As in the previous examples, the statistic keywords are located in the column definition.

```
PROC TABULATE DATA=TEMP;
   VAR INCOME;
   CLASS EDUC;
   TABLE EDUC, INCOME*N INCOME*MEAN INCOME*MAX;
RUN;
```

As you can see in Output 5.4, this produces a table with three rows and three columns. Notice that the labels for N, MEAN, and MAX are in the column headings. PROC TABULATE puts the statistic labels in the column heading because the statistics were requested as part of the column definition.

But you don't have to do it this way. You can also request statistics in the row definition. The following code will produce the same values in the table cells as in the previous table.

```
PROC TABULATE DATA=TEMP;
   VAR INCOME;
   CLASS EDUC;
   TABLE EDUC*N EDUC*MEAN EDUC*MAX, INCOME;
RUN;
```

Now the series of statistics are added to the rows of the table in Output 5.5. The first group of rows shows the N for each value of education, the next group shows the MEAN, and the final group shows the MAX. If you look at the values, you can see that although the format of the table has changed, the values have not. PROC TABULATE gives you a lot of flexibility to reformat your table whatever way you want. You can use this flexibility to improve the readability of your table.

That said, there is one important limitation to where you can place your statistic definitions. You can't list statistics in both the row definition and the column definition.

```
PROC TABULATE DATA=TEMP;
   VAR INCOME;
   CLASS EDUC;
   TABLE EDUC*N, INCOME*MEAN INCOME*MAX;
RUN;
```

This code produces the error message shown in Output 5.6. This is because a two-dimensional table is a cross-tabulation of the row variables and column variables. PROC TABULATE can figure out what to do if you ask for EDUC by N, EDUC by MEAN, or INCOME by N, but it doesn't know what to do with N by MEAN or N by MAX.

Output 5.4

```
---------------------------------------------------------------------
|                             | Income  | Income   | Income   |
|                             |---------+----------+----------|
|                             |   N     |  MEAN    |  MAX     |
|-----------------------------+---------+----------+----------|
|Education                    |         |          |          |
|-----------------------------|         |          |          |
|<HS                          | 1001.00 | 11937.52 |  87130.00|
|-----------------------------+---------+----------+----------|
|HS                           | 2282.00 | 19581.80 | 133567.00|
|-----------------------------+---------+----------+----------|
|College                      | 3356.00 | 32710.58 | 263253.00|
---------------------------------------------------------------------
```

Output 5.5

```
-----------------------------------------------------------
|                                    |   Income   |
|------------------------------------+------------|
|Education        |                  |            |
|-----------------+----------------- |            |
|<HS              |N                 |    1001.00 |
|-----------------+--------+---------+----------- |
|HS               |N                 |    2282.00 |
|-----------------+--------+---------+----------- |
|College          |N                 |    3356.00 |
|-----------------+--------+---------+----------- |
|Education        |                  |            |
|-----------------+----------------- |            |
|<HS              |MEAN              |   11937.52 |
|-----------------+--------+---------+----------- |
|HS               |MEAN              |   19581.80 |
|-----------------+--------+---------+----------- |
|College          |MEAN              |   32710.58 |
|-----------------+--------+---------+----------- |
|Education        |                  |            |
|-----------------+----------------- |            |
|<HS              |MAX               |   87130.00 |
|-----------------+--------+---------+----------- |
|HS               |MAX               |  133567.00 |
|-----------------+--------+---------+----------- |
|College          |MAX               |  263253.00 |
-----------------------------------------------------------
```

Output 5.6

```
48    PROC TABULATE DATA=TEMP;
49         VAR INCOME;
50         CLASS EDUC;
51         TABLE EDUC*N, INCOME*MEAN INCOME*MAX;
52    RUN;

ERROR: There are multiple statistics associated with a single table cell in the following nesting :
EDUC * N * INCOME * MEAN.
ERROR: There are multiple statistics associated with a single table cell in the following nesting :
EDUC * N * INCOME * MAX.
NOTE: The SAS System stopped processing this step because of errors.
```

Adding Classification Variables on Both Dimensions

The previous example shows how to reformat a table from one to two dimensions, but it does not show the true power of two-dimensional tables. With two dimensions, you can classify your statistics by two different variables at the same time.

To do this, you put one classification variable in the row dimension and one classification in the column dimension. The previous example has education displayed in rows and income is the column variable. In this new table, we will add employment status as an additional column variable. Instead of displaying only the N and MEAN of the variable INCOME, we will display these statistics broken down by part-time versus full-time employment status.

So in the following code, we leave EDUC as the row variable, and we leave INCOME in the column dimension. The only change is to add FULLTIME to the column dimension using the asterisk operator. This tells PROC TABULATE to break down each of the column elements into categories of employment status.

```
PROC TABULATE DATA=TEMP;
   VAR INCOME;
   CLASS EDUC FULLTIME;
   TABLE EDUC, INCOME*FULLTIME*N INCOME*FULLTIME*MEAN;
RUN;
```

This table is shown in Output 5.7. Notice how the analysis variable INCOME remains as the column heading, and N and MEAN remain as the statistics, but now there are additional column headings to show the two categories of FULLTIME.

With this many headings, the table can be a bit hard to read. Also, the TABLE statement has become rather lengthy. To simplify both the table and the TABLE statement, we can use parentheses in the TABLE statement.

```
PROC TABULATE DATA=TEMP;
   VAR INCOME;
   CLASS EDUC FULLTIME;
   TABLE EDUC, INCOME*FULLTIME*(N MEAN);
RUN;
```

In Output 5.8, you can see that the contents of the table cells did not change, they're just rearranged. Now the N and MEAN for each category of FULLTIME are grouped together. Also, the INCOME and FULLTIME headings span all four columns, making the table easier to read. With PROC TABULATE, you can produce the same table cells by using a variety of TABLE statements. For example, if you'd rather have the two N statistics and the two means side-by-side, you can move the parentheses as follows. The results are shown in Output 5.9.

```
PROC TABULATE DATA=TEMP;
   VAR INCOME;
   CLASS EDUC FULLTIME;
   TABLE EDUC, INCOME*(N MEAN)*FULLTIME;
RUN;
```

Output 5.7

	Income		Income	
	Employment		Employment	
	Full-time	Part-time	Full-time	Part-time
	N	N	MEAN	MEAN
Education				
<HS	257.00	44.00	19267.53	19755.32
HS	1198.00	165.00	24798.92	23363.48
College	2257.00	271.00	38179.60	35092.72

Output 5.8

	Income			
	Employment			
	Full-time		Part-time	
	N	MEAN	N	MEAN
Education				
<HS	257.00	19267.53	44.00	19755.32
HS	1198.00	24798.92	165.00	23363.48
College	2257.00	38179.60	271.00	35092.72

Output 5.9

	Income			
	N		MEAN	
	Employment		Employment	
	Full-time	Part-time	Full-time	Part-time
Education				
<HS	257.00	44.00	19267.53	19755.32
HS	1198.00	165.00	24798.92	23363.48
College	2257.00	271.00	38179.60	35092.72

Adding Another Classification Variable

The previous example showed how to add a classification variable to both the rows and columns of a two-dimensional table. But you are not limited to just one classification per dimension. This example shows how to display additional subgroups of the data.

In this case, we're going to add gender as an additional row classification. The variable is added to the CLASS statement and to the row dimension of the TABLE statement.

```
PROC TABULATE DATA=TEMP;
   VAR INCOME;
   CLASS EDUC FULLTIME GENDER;
   TABLE EDUC GENDER, INCOME*FULLTIME*MEAN;
RUN;
```

In the results shown in Output 5.10, you can see that we now have two two-dimensional "mini-tables" within a single table. First, we have a table of income by employment status and education, and then we have a table showing income by employment status and gender.

This ability to stack multiple mini-tables within a single table can be a powerful tool for delivering large quantities of information in a user-friendly format. For example, if you want to show your manager the results of a survey on customer satisfaction, it would be helpful to be able to break down the results by a number of demographic variables in a single table.

The following code generates a table that makes it easy to spot the pockets of dissatisfied customers so that your company can better target its customer service efforts. The code is set up exactly like the preceding table, except that there are many more class variables in the row dimension.

```
PROC TABULATE DATA=TEMP;
   CLASS GENDER RACE CUSTTYPE SIZE;
   VAR SAT;
   TABLE GENDER RACE CUSTTYPE SIZE, SAT*(N MEAN);
RUN;
```

From the output shown in Output 5.11, you can see that we now have a table made up of four mini-tables. This allows a detailed examination of customer satisfaction broken down into a number of categories. You can easily see that satisfaction is roughly equal in all subcategories but one: females are noticeably less satisfied than males. Because more than half of your customers are female, this is a major problem for your company to address.

Using one simple PROC TABULATE table, you have clearly summarized a large quantity of information in an easily understood format. For your manager, this is important information that will enable efficient targeting of resources to improve overall customer satisfaction.

Output 5.10

	Income	
	Employment	
	Full-time	Part-time
	MEAN	MEAN
Education		
<HS	19267.53	19755.32
HS	24798.92	23363.48
College	38179.60	35092.72
Gender		
Female	23821.25	24048.34
Male	40656.44	35598.25

Output 5.11

	Satisfaction Rating	
	N	MEAN
Gender		
Female	3450.00	3.03
Male	2991.00	5.04
Race		
White	6261.00	3.96
Black	165.00	4.02
Other	15.00	3.93
Custtype		
Retail	3263.00	3.94
Wholesale	3178.00	3.99
Size		
Large	2654.00	3.92
Small	3787.00	3.99

Nesting the Classification Variables

So far, all we have done is added additional "tables" to the bottom of our first table. We used the space operator between each of the row variables to produce concatenated tables.

By using a series of row variables, we could explore a variety of relationships between the variables. In the customer satisfaction table, we could see how satisfaction varies by gender, and we could see how satisfaction varies by race, but we could not see how gender and race interacted. And in the income table, we could see how income varied by education and by gender, but not how the two variables interact to affect income.

But we can learn even more from our data using PROC TABULATE. The power of PROC TABULATE comes from being able to look at *combinations* of categories within a single table. In the following example, we build a table to look at income by employment status for combinations of education and gender.

This code is the same as we used for the first table in the previous example. The only change is that in the row definition, the asterisk operator is used to show that we want to *nest* the two row variables. In other words, we want to see the breakdown of income by gender within each category of education.

```
PROC TABULATE DATA=TEMP;
   VAR INCOME;
   CLASS EDUC FULLTIME GENDER;
   TABLE EDUC*GENDER, INCOME*FULLTIME*MEAN;
RUN;
```

As you can see in Output 5.12, this code produces nested categories within the row headings. The row headings are now split into two columns. The first column shows education and the second column shows gender. It is now possible to interpret the interaction of gender and education.

You can also reverse the order of the row variables to look at education within gender, instead of gender within education. All you do is move GENDER so that it comes before EDUC. PROC TABULATE always produces the nested rows in the order the variables are listed on the TABLE statement.

```
PROC TABULATE DATA=TEMP;
   VAR INCOME;
   CLASS EDUC FULLTIME GENDER;
   TABLE GENDER*EDUC, INCOME*FULLTIME*MEAN;
RUN;
```

This table is shown in Output 5.13. This table has the same cells as the previous table, but they have been rearranged. PROC TABULATE gives you the control to vary the organization of your tables so that you can learn as much from your data as possible. You should try configuring your tables in many different ways to see what story emerges.

Output 5.12

```
---------------------------------------------------------------
|                              |        Income              |
|                              |--------------------------- |
|                              |       Employment           |
|                              |--------------------------- |
|                              | Full-time  | Part-time     |
|                              |------------+-------------   |
|                              |    MEAN    |    MEAN        |
|------------------------------+------------+-------------   |
|Education      |Gender        |            |               |
|---------------+--------------|            |               |
|<HS            |Female        |   14850.22 |   14449.73    |
|               |--------------+------------+-------------   |
|               |Male          |   22625.90 |   25060.91    |
|---------------+--------------+------------+-------------   |
|HS             |Female        |   18189.30 |   17574.26    |
|               |--------------+------------+-------------   |
|               |Male          |   31085.59 |   28307.08    |
|---------------+--------------+------------+-------------   |
|College        |Female        |   27745.10 |   28767.81    |
|               |--------------+------------+-------------   |
|               |Male          |   47960.27 |   42817.40    |
---------------------------------------------------------------
```

Output 5.13

```
---------------------------------------------------------------
|                              |        Income              |
|                              |--------------------------- |
|                              |       Employment           |
|                              |--------------------------- |
|                              | Full-time  | Part-time     |
|                              |------------+-------------   |
|                              |    MEAN    |    MEAN        |
|------------------------------+------------+-------------   |
|Gender         |Education     |            |               |
|---------------+--------------|            |               |
|Female         |<HS           |   14850.22 |   14449.73    |
|               |--------------+------------+-------------   |
|               |HS            |   18189.30 |   17574.26    |
|               |--------------+------------+-------------   |
|               |College       |   27745.10 |   28767.81    |
|---------------+--------------+------------+-------------   |
|Male           |<HS           |   22625.90 |   25060.91    |
|               |--------------+------------+-------------   |
|               |HS            |   31085.59 |   28307.08    |
|               |--------------+------------+-------------   |
|               |College       |   47960.27 |   42817.40    |
---------------------------------------------------------------
```

Adding Statistics to the Nested Variables

The previous examples have showed how to add multiple classification variables to your two-dimensional table. But often you want to add multiple statistics as well. So far, we have looked at means, but the tables would be more helpful if we could also see the number of observations for each subgroup. If there are only a few observations in some of the table cells, then the mean statistic becomes less useful.

The following code produces a table with the number of observations next to every mean. This is accomplished by using parentheses around the two statistics keywords N and MEAN.

```
PROC TABULATE DATA=TEMP;
   VAR INCOME;
   CLASS EDUC FULLTIME GENDER;
   TABLE EDUC*GENDER*(N MEAN), INCOME*FULLTIME;
RUN;
```

With this code, you get a table that shows two rows of statistics for each category, as shown in Output 5.14. You can see by looking at the values of N that there are very few observations for people with less than a high-school education who are working part-time. Based on this information, you can decide to recategorize your education variable to make the table more meaningful.

The following code uses a recoded education variable called EDUC2 that only has two categories: "<=High School" and "> High School." The other change that is made in this code is to move the statistics keywords N and MEAN to the column definition. This improves the readability of the table.

```
PROC TABULATE DATA=TEMP;
   VAR INCOME;
   CLASS EDUC2 FULLTIME GENDER;
   TABLE EDUC2*GENDER, INCOME*FULLTIME*(N MEAN);
RUN;
```

Now, with only two categories of education, each cell in Output 5.15 has a larger number of observations. Also, the table is much easier to read. In the first table, the N statistics and MEAN statistics alternated as you moved down each column, making it hard to compare the various N statistics. Now the N statistics are grouped in a single column, making it easy to read down the table and compare the values. The means, too, are grouped in a single column.

Another advantage to this formatting is that the table is much more compact. Part of this is due to the removal of one of the categories of education, but space was also saved by moving the statistic labels to the column headings where they do not have to be repeated for each classification subgroup.

As you build tables with multiple CLASS variables, it will become increasingly important to think about where to place your statistics. If the row or column headings end up having four or five stacked or nested labels, your table will be nearly impossible for end users to interpret.

Output 5.14

```
----------------------------------------------------------------
|                              |            Income              |
|                              |-------------------------------|
|                              |         Employment            |
|                              |-------------------------------|
|                              | Full-time  | Part-time  |
|------------------------------+------------+-----------|
|Education|Gender  |           |            |           |
|---------+--------+-----------|            |           |
|<HS      |Female  |N          |    111.00  |    22.00  |
|         |        |-----------+------------+-----------|
|         |        |MEAN       |  14850.22  | 14449.73  |
|         |--------+-----------+------------+-----------|
|         |Male    |N          |    146.00  |    22.00  |
|         |        |-----------+------------+-----------|
|         |        |MEAN       |  22625.90  | 25060.91  |
|---------+--------+-----------+------------+-----------|
|HS       |Female  |N          |    584.00  |    76.00  |
|         |        |-----------+------------+-----------|
|         |        |MEAN       |  18189.30  | 17574.26  |
|         |--------+-----------+------------+-----------|
|         |Male    |N          |    614.00  |    89.00  |
|         |        |-----------+------------+-----------|
|         |        |MEAN       |  31085.59  | 28307.08  |
|---------+--------+-----------+------------+-----------|
|College  |Female  |N          |   1092.00  |   149.00  |
|         |        |-----------+------------+-----------|
|         |        |MEAN       |  27745.10  | 28767.81  |
|         |--------+-----------+------------+-----------|
|         |Male    |N          |   1165.00  |   122.00  |
|         |        |-----------+------------+-----------|
|         |        |MEAN       |  47960.27  | 42817.40  |
----------------------------------------------------------------
```

Output 5.15

		Income			
		Employment			
		Full-time		Part-time	
		N	MEAN	N	MEAN
EDUC2	Gender				
<= High School	Female	695.00	17656.01	98.00	16872.84
	Male	760.00	29460.44	111.00	27663.69
> High School	Female	1092.00	27745.10	149.00	28767.81
	Male	1165.00	47960.27	122.00	42817.40

Adding Totals to the Rows and Columns

One trick that will definitely make your tables more readable is row and column totals. Totals are quite easy to generate in PROC TABULATE.

For example, take a look at the following code, which does not ask for totals:

```
PROC TABULATE DATA=TEMP;
    CLASS EDUC2 GENDER FULLTIME;
    TABLE EDUC2*GENDER, FULLTIME*N;
RUN;
```

This produces the table shown in Output 5.16. This is a nice table, but it shows only the details. You can see that there are 98 women with education "<= High School" that are working part-time. But if you want to see how this compares to the total number of women in this row (both those who work part-time and those who work full-time) you have to do some math. By adding 98 to 695, you can figure out that 98 out of 793 of these women work part-time. It would be easier to read the table if it had row totals.

The following code will produce the same table as before, but with the addition of row totals:

```
PROC TABULATE DATA=TEMP;
    CLASS EDUC2 GENDER FULLTIME;
    TABLE EDUC2*GENDER, FULLTIME*N ALL;
RUN;
```

As you can see from Output 5.17, the table now has row totals. But you're probably wondering about the code that we just used. Where did this variable ALL come from? The answer is that PROC TABULATE provided it. ALL is a built-in classification variable available from PROC TABULATE. You do not have to list it in the CLASS statement because it is a classification variable by definition. You can just use ALL in your TABLE statements as needed.

In this case, ALL was added as a column total by putting the ALL keyword at the end of the TABLE statement. It is separated from the rest of the column dimension by a space because it is to be a new column, not a new classification of an existing column.

Not only can you use ALL to add row totals, but you can also use ALL to produce column totals. What you do is list ALL as an additional variable in the row definition of the TABLE statement. No asterisk is needed because we just want to add a total at the bottom of the table.

```
PROC TABULATE DATA=TEMP;
    CLASS EDUC2 GENDER FULLTIME;
    TABLE EDUC2*GENDER ALL, FULLTIME*N;
RUN;
```

The resulting table is shown in Output 5.18. Now we can see the overall totals for each employment status.

Output 5.16

		Employment	
		Full-time	Part-time
		N	N
EDUC2	Gender		
<= High School	Female	695.00	98.00
	Male	760.00	111.00
> High School	Female	1092.00	149.00
	Male	1165.00	122.00

Output 5.17

		Employment		
		Full-time	Part-time	ALL
		N	N	N
EDUC2	Gender			
<= High School	Female	695.00	98.00	793.00
	Male	760.00	111.00	871.00
> High School	Female	1092.00	149.00	1241.00
	Male	1165.00	122.00	1287.00

Output 5.18

		Employment	
		Full-time	Part-time
		N	N
EDUC2	Gender		
<= High School	Female	695.00	98.00
	Male	760.00	111.00
> High School	Female	1092.00	149.00
	Male	1165.00	122.00
ALL		3712.00	480.00

Adding Subtotals

ALL can do a lot more for your tables than just add totals. You can also use it to create subtotals when you have nested classification variables. The following examples show how you can create a variety of different subtotals for the same table.

This is the original code. This produces the basic table that we've been working with throughout the chapter. It has nested row headings and a single column variable.

```
PROC TABULATE DATA=TEMP;
   CLASS EDUC2 GENDER FULLTIME;
   TABLE EDUC2*GENDER, FULLTIME*N;
RUN;
```

The next examples show two of the many ways you can add subtotals to this table. The first method puts subtotals for each gender at the bottom of the table. The resulting table is shown in Output 5.19.

```
PROC TABULATE DATA=TEMP;
   CLASS EDUC2 GENDER FULLTIME;
   TABLE (EDUC2 ALL)*GENDER, FULLTIME*N;
RUN;
```

To understand how this works, expand the row definition: "(EDUC2 ALL)*GENDER" is equivalent to "EDUC2*GENDER ALL*GENDER", so you get two concatenated tables, one of education by gender and one of subtotals by gender.

The second example shows how to add subtotals for each level of education. The resulting table is shown in Output 5.20.

```
PROC TABULATE DATA=TEMP;
   CLASS EDUC2 GENDER FULLTIME;
   TABLE EDUC2*(GENDER ALL), FULLTIME*N;
RUN;
```

If you expand the row definition for this example, "EDUC2*(GENDER ALL)" is equivalent to "EDUC2*GENDER EDUC2*ALL", so for each level of education, you get a table showing the breakdown by gender and then a subtotal for both genders. In this case, the subtotals for education are not grouped together at the bottom of the table. This is because education is the outer row variable (the highest level of nesting).

If you wanted to put these subtotals together at the bottom of the table, you would have to reverse the order of the row variables, as in the following code (output not shown);

```
PROC TABULATE DATA=TEMP;
   CLASS EDUC2 GENDER FULLTIME;
   TABLE (GENDER ALL)*EDUC2, FULLTIME*N;
RUN;
```

Output 5.19

```
-------------------------------------------------------------------
|                               |        Employment            |
|                               |------------------------------|
|                               | Full-time   | Part-time      |
|                               |-------------+--------------- |
|                               |     N       |      N         |
|-------------------------------+-------------+--------------- |
|EDUC2          |Gender         |             |                |
|---------------+---------------|             |                |
|<= High School |Female         |      695.00 |        98.00   |
|               |---------------+-------------+--------------- |
|               |Male           |      760.00 |       111.00   |
|---------------+---------------+-------------+--------------- |
|> High School  |Female         |     1092.00 |       149.00   |
|               |---------------+-------------+--------------- |
|               |Male           |     1165.00 |       122.00   |
|---------------+---------------+-------------+--------------- |
|ALL            |Female         |     1787.00 |       247.00   |
|               |---------------+-------------+--------------- |
|               |Male           |     1925.00 |       233.00   |
-------------------------------------------------------------------
```

Output 5.20

```
-------------------------------------------------------------------
|                               |        Employment            |
|                               |------------------------------|
|                               | Full-time   | Part-time      |
|                               |-------------+--------------- |
|                               |     N       |      N         |
|-------------------------------+-------------+--------------- |
|EDUC2          |Gender         |             |                |
|---------------+---------------|             |                |
|<= High School |Female         |      695.00 |        98.00   |
|               |---------------+-------------+--------------- |
|               |Male           |      760.00 |       111.00   |
|               |---------------+-------------+--------------- |
|               |ALL            |     1455.00 |       209.00   |
|---------------+---------------+-------------+--------------- |
|> High School  |Gender         |             |                |
|               |---------------|             |                |
|               |Female         |     1092.00 |       149.00   |
|               |---------------+-------------+--------------- |
|               |Male           |     1165.00 |       122.00   |
|               |---------------+-------------+--------------- |
|               |ALL            |     2257.00 |       271.00   |
-------------------------------------------------------------------
```

Putting It All Together

This final example pulls together each of the PROC TABULATE features demonstrated in this chapter. This will be a two-dimensional table with multiple classification variables, column subtotals, and row and column totals. While this is a complex table, you can break it down into each of the elements that have already been presented.

The starting point for the table is the two-dimensional table with column subtotals from the previous example (this table is shown in Output 5.20). The only change made to this code is to switch the N statistic to a MEAN and add INCOME as an analysis variable.

```
PROC TABULATE DATA=TEMP;
    CLASS EDUC2 GENDER FULLTIME;
    VAR INCOME;
    TABLE EDUC2*(GENDER ALL), FULLTIME*INCOME*MEAN;
RUN;
```

Because this table is virtually the same as in the previous example, the output is not shown. For this example, we are going to add a total to the bottom of each column, in addition to the subtotals that are already included for each level of education. To do this, we add the variable ALL to the row dimension a second time.

```
PROC TABULATE DATA=TEMP;
    CLASS EDUC2 GENDER FULLTIME;
    VAR INCOME;
    TABLE EDUC2*(GENDER ALL) ALL, FULLTIME*INCOME*MEAN;
RUN;
```

The resulting table is shown in Output 5.21. Notice that because the second reference to ALL is not combined with any of the other row variables, this variable is added to the bottom of the table by itself. The final step to building our table is to add the row totals. This is done just as in the prior example, by adding the variable ALL to the column dimension.

```
PROC TABULATE DATA=TEMP;
    CLASS EDUC2 GENDER FULLTIME;
    VAR INCOME;
    TABLE EDUC2*(GENDER ALL) ALL, (FULLTIME ALL)*INCOME*MEAN;
RUN;
```

The final table is shown in Output 5.22. There's a lot going on in this table. If you don't understand how this TABLE statement works, stop and take the time to break it down.

In this case, the row dimension defines three mini-tables: the first table is EDUC2*GENDER, the second is EDUC2*ALL, and the third is the total ALL. What makes it a bit confusing is that the items from the second mini-table are mixed in to the nestings of the first mini-table: the totals for each education level follow the gender breakdown for that level. The column dimension is easier. It has two parts: the first is FULLTIME*INCOME*MEAN or the mean income for each employment status; the second part is ALL*INCOME*MEAN or the overall mean income. Anytime you find a table confusing, be sure to take the time to break it down and be sure you know how it works.

Output 5.21

		Employment	
		Full-time	Part-time
		Income	Income
		MEAN	MEAN
EDUC2	Gender		
<= High School	Female	17656.01	16872.84
	Male	29460.44	27663.69
	ALL	23821.90	22603.87
> High School	Gender		
	Female	27745.10	28767.81
	Male	47960.27	42817.40
	ALL	38179.60	35092.72
ALL		32551.78	29654.86

Output 5.22

		Employment		
		Full-time	Part-time	ALL
		Income	Income	Income
		MEAN	MEAN	MEAN
EDUC2	Gender			
<= High School	Female	17656.01	16872.84	17559.22
	Male	29460.44	27663.69	29231.46
	ALL	23821.90	22603.87	23668.91
> High School	Gender			
	Female	27745.10	28767.81	27867.89
	Male	47960.27	42817.40	47472.76
	ALL	38179.60	35092.72	37848.69
ALL		32551.78	29654.86	32220.08

Creating Three-Dimensional Tables

Now that you have mastered two-dimensional tables, let's add a third dimension. You may be asking yourself: Three dimensions? How do you print a table shaped like a cube?

Actually, a three-dimensional table is not shaped like a cube, it looks just like a two-dimensional table, except that it spans multiple pages. A one-dimensional table has columns. A two-dimensional table has both columns and rows. A three-dimensional table is a two-dimensional table that is repeated across multiple pages. Basically, you print a new page for each value of the page variable.

The Example

Once again we are looking at mean income. In this case, we want to see how it varies by gender, occupation, and employment sector. It would be easy to put any two of these three classifications in a table. But to put all three on one page would be difficult both to fit and to interpret. This is where the page dimension comes in handy. We are going to build a table of income by gender and occupation, and then repeat the table across multiple pages, one for each employment sector.

Sector=xxx	Income		
	Male	Female	ALL
	MEAN	MEAN	MEAN
Occupation			
Managerial	999	999	999
Sales	999	999	999
Technical	999	999	999

Start with Building the Two-Dimensional Table

The biggest challenge in building three-dimensional tables is making sense of the TABLE statement. So the best way to start is with a two-dimensional table. Once you've got that set up correctly, it's relatively easy to add the page variable to expand the table to multiple pages.

For our example, we're going to build a table of income by occupation and gender, and then we're going to add sector as the page variable. We'll end up with two pages of tables, the first page will have the first sector (Private), and the second page will have the second sector (Public).

Ignoring the third dimension for now, let's build the basic table. This table has rows showing occupation and columns showing income by gender. The code is as follows:

```
PROC TABULATE DATA=TEMP;
    CLASS OCCUP GENDER;
    VAR INCOME;
    TABLE OCCUP, (GENDER ALL)*INCOME*MEAN;
RUN;
```

This produces the table shown in Output 6.1. At this point, you want to look the table over carefully and be sure that you've got exactly what you want to see in your final table. The only difference between this table and our final three-dimensional table is that right now, the table is showing results for both SECTOR="Private" and SECTOR="Public" combined. In the final tables, each sector will be shown on its own page.

Because the two-dimensional table looks correct, we'll go on to adding the third dimension. Just like when we converted a one-dimensional table to a two-dimensional table, we add a new dimension with a comma operator in the TABLE statement. We add the new variable SECTOR to the existing TABLE statement with a comma to separate it from the row and column definitions.

You might think that the following code is the correct way to add the third dimension to the TABLE statement. However, what this would generate is a table with GENDER as the rows, SECTOR as the columns, and OCCUP as the pages!

```
TABLE OCCUP, (GENDER ALL)*INCOME*MEAN, SECTOR;
```

To add the third dimension to a table correctly, you add it at the *beginning* of the TABLE statement. Remember that this was also true when we added the second dimension to the table. We added rows to the columns by adding a row definition before the column definition. The correct code for our table is as follows:

```
PROC TABULATE DATA=TEMP;
    CLASS OCCUP GENDER SECTOR;
    VAR INCOME;
    TABLE SECTOR, OCCUP, (GENDER ALL)*INCOME*MEAN;
RUN;
```

The resulting tables are shown in Output 6.2. To save space, both tables are shown on a single page. In reality, PROC TABULATE puts a page break between each of the tables, and the table for SECTOR="Public" would appear on a second page. Notice that each table is automatically given a title indicating the sector it represents.

Output 6.1

	Gender		
	Male	Female	ALL
	Income	Income	Income
	MEAN	MEAN	MEAN
Occupation			
Managerial	55818.31	32128.29	45757.74
Professional	50055.37	32364.57	40463.48
Technical	34680.32	25386.12	29968.68

Output 6.2

Sector Private

	Gender		
	Male	Female	ALL
	Income	Income	Income
	MEAN	MEAN	MEAN
Occupation			
Managerial	55948.92	32337.14	46075.54
Professional	51926.11	31693.72	42098.03
Technical	35336.78	25162.63	30209.65

Sector Public

	Gender		
	Male	Female	ALL
	Income	Income	Income
	MEAN	MEAN	MEAN
Occupation			
Managerial	54836.21	32044.57	44390.04
Professional	41492.10	33635.86	36029.82
Technical	29510.75	26975.44	28168.53

Comparison to Using a BY Statement

There is another way to create a three-dimensional table. Like many SAS procedures, PROC TABULATE supports the BY statement. You can use this to repeat a table for each value of the BY variable.

For example, the following table can be built by using a three-dimensional TABLE statement or by a two-dimensional TABLE statement and a BY variable.

The following code takes the first approach. It uses a three-dimensional TABLE statement and produces the table shown in Output 6.3. Reading the TABLE statement from right to left, it calls for columns showing the classifications GENDER and ALL, a single row showing MEAN income, and it is repeated for each value of RACE.

```
PROC TABULATE DATA=TEMP;
   CLASS RACE GENDER;
   VAR INCOME;
   TABLE RACE, INCOME*MEAN, GENDER ALL;
RUN;
```

You can get the same tables with a simpler TABLE statement. By moving RACE from the page dimension of the TABLE statement to a BY statement, you end up with the same tables produced by a simple two-dimensional TABLE statement. To use this approach, you have to sort the data by the BY variable, in this case RACE.

```
PROC SORT DATA=TEMP;
   BY RACE;
RUN;
```

Then you can run the following PROC TABULATE code:

```
PROC TABULATE DATA=TEMP;
   BY RACE;
   CLASS RACE GENDER;
   VAR INCOME;
   TABLE INCOME*MEAN, GENDER ALL;
RUN;
```

The resulting tables are shown in Output 6.4. The only difference between these tables and the previous tables in Output 6.3 is the format of the titles. The tables produced with the BY statement have an "=" between the variable name and the classification value.

So if you find three-dimensional tables confusing, this is one way to simplify them. However, there are some costs. First, running the PROC SORT before the PROC TABULATE is far less efficient than running the three-dimensional PROC TABULATE code alone. Second, you have a lot less control over the placement of the page variable labels in the title. As you will see later in this chapter, PROC TABULATE gives you the ability to modify the placement of these labels. Finally, PROC TABULATE lets you construct more complex page variables, including the ability to display a table for the totals as well as for each value of the page variable.

Output 6.3

Race White

```
-----------------------------------------------------------------
|                               |          Gender          |       |
|                               |--------------------------|       |
|                               |   Male   |  Female  |    ALL    |
|-------------------------------+----------+----------+-----------|
|Income         |MEAN           |  34290.05|  18022.33|  25344.29 |
-----------------------------------------------------------------
```

Race Black

```
-----------------------------------------------------------------
|                               |          Gender          |       |
|                               |--------------------------|       |
|                               |   Male   |  Female  |    ALL    |
|-------------------------------+----------+----------+-----------|
|Income         |MEAN           |  26676.78|  13460.62|  19460.60 |
-----------------------------------------------------------------
```

Output 6.4

Race=White

```
-----------------------------------------------------------------
|                               |          Gender          |       |
|                               |--------------------------|       |
|                               |   Male   |  Female  |    ALL    |
|-------------------------------+----------+----------+-----------|
|Income         |MEAN           |  34290.05|  18022.33|  25344.29 |
-----------------------------------------------------------------
```

Race=Black

```
-----------------------------------------------------------------
|                               |          Gender          |       |
|                               |--------------------------|       |
|                               |   Male   |  Female  |    ALL    |
|-------------------------------+----------+----------+-----------|
|Income         |MEAN           |  26676.78|  13460.62|  19460.60 |
-----------------------------------------------------------------
```

More Complex Page Dimensions: Adding a Page for the Totals

So far in this chapter, we've built three-dimensional tables with a single page variable. However, there's no rule that says you can have only one variable in the page dimension. In fact, you can put anything in the page dimension that you can put in a row or column dimension.

For example, if you really wanted to, you could produce a table with the following code:

```
PROC TABULATE DATA=TEMP;
    CLASS RACE GENDER EDUC FULLTIME SECTOR;
    VAR INCOME;
    TABLE SECTOR*RACE*EDUC, FULLTIME, GENDER*INCOME*(MEAN MIN MAX);
RUN;
```

This table is not shown because it would take up most of the rest of this book. What you would get would be a table of FULLTIME by INCOME for every combination of SECTOR, RACE, and EDUC. Depending on how the variables are formatted, this could generate dozens of pages of tables.

Complex tables with multiple class variables spread over many pages are extremely hard to understand. While there may be times when you want to do this, in general it is better to keep the page dimension simple and put the more complex parts of the table in the row and column dimensions.

For example, the previous table would be a little clearer if it we reversed the page and row dimensions:

```
PROC TABULATE DATA=TEMP;
    CLASS RACE GENDER EDUC FULLTIME SECTOR;
    VAR INCOME;
    TABLE FULLTIME, SECTOR*RACE*EDUC, GENDER*INCOME*(MEAN MIN MAX);
RUN;
```

This table may still be too complex, but at least now the complex part is contained on each page. The page dimension is a simple breakdown by FULLTIME.

The best way to use the page dimension is for repeating tables for the classifications of a single variable. However, there is one enhancement that you may want to consider: adding a page for the totals. If you were producing a series of tables for each level of education, it would be nice to end with a table showing the totals for all levels of education, as in the following example:

```
PROC TABULATE DATA=TEMP;
    CLASS EDUC RACE GENDER;
    VAR INCOME;
    TABLE EDUC ALL, RACE*GENDER, INCOME*(MEAN MIN MAX);
RUN;
```

All you have to do to get a final page with the totals is add the CLASS variable ALL to the page dimension. By using a space as the operator, you tell PROC TABULATE that you want a series of tables for each level of education, followed by a single table for ALL. The results are shown in Output 6.5 (the page breaks have been removed to save space).

Output 6.5

Education HS

		Income		
		MEAN	MIN	MAX
Race	Gender			
White	Male	26627.88	0.00	120759.00
	Female	13715.56	0.00	106759.00
Black	Male	18506.05	2820.00	44693.00
	Female	8643.96	0.00	28877.00

Education College

		Income		
		MEAN	MIN	MAX
Race	Gender			
White	Male	43899.85	0.00	251998.00
	Female	23681.98	0.00	263253.00
Black	Male	33986.82	0.00	118000.00
	Female	20491.55	0.00	52900.00

ALL

		Income		
		MEAN	MIN	MAX
Race	GENDER			
White	Male	37419.14	0.00	251998.00
	Female	19560.98	0.00	263253.00
Black	Male	29149.08	0.00	118000.00
	Female	15790.13	0.00	52900.00

Putting a Title in Each Table

You've probably noticed that most PROC TABULATE tables have a big empty box in the table grid. The box sits above the row headers and to the left of the column headers. Wouldn't it be nice if we didn't have to waste this space? As it happens, you can take advantage of that space. You can use that space to hold some descriptive text or the value of some variable. But the real advantage of this space is that for three-dimensional tables, you can use it to hold the value of the page variable(s).

First let's look at a simpler two-dimensional example. We are going to add a new feature to our TABLE statement. PROC TABULATE supports a number of optional features for the TABLE statement. To specify one of the options, you follow the TABLE statement with a slash and then specify the option setting. In this case, we're going to specify the BOX= option. This option tells PROC TABULATE that we want to put something in the empty box at the top left corner of our table.

One thing you can put in the box is a label. Suppose that we want to label our table with the source of the data. This code tells PROC TABULATE to put the literal string "Customer Survey Data" in the box.

```
PROC TABULATE DATA=TEMP;
   CLASS RACE GENDER;
   VAR SAT;
   TABLE RACE, GENDER*SAT
      / BOX="Customer Survey Data";
RUN;
```

You can see the results in Output 6.6. You can also specify BOX= using a variable. If we revise the code with the following TABLE statement, PROC TABULATE will use the variable SAT to create the text for the box.

```
TABLE RACE, GENDER*SAT
   / BOX=SAT;
```

If SAT has a label, then PROC TABULATE will put that label in the box. Otherwise, PROC TABULATE will put the name of the variable in the box. As you can see from Output 6.7, SAT has the label "Satisfaction Rating," so that is what is displayed in the box. This doesn't seem very useful right now because this label is also displayed in the column headings, but its use will become more apparent in later chapters.

The best use of the BOX= option is for three-dimensional tables. That's because there's a special variable you can use with the option to put the value of the page variable(s) inside the box. If you specify BOX=_PAGE_, then PROC TABULATE will take the page labels that used to appear as titles above the table and put them inside the table in the box. This makes the table easier to read, and it is a handy space saver.

```
PROC TABULATE DATA=TEMP;
   CLASS RACE GENDER CUSTTYPE;
   VAR SAT;
   TABLE CUSTTYPE, RACE, GENDER*SAT
      / BOX=_PAGE_;
RUN;
```

The results are shown in Output 6.8 (only the first page is shown).

Output 6.6

Customer Survey Data	Gender	
	Male	Female
	Satisfaction Rating	Satisfaction Rating
	SUM	SUM
Race		
White	13600.00	10152.00
Black	375.00	251.00

Output 6.7

Satisfaction Rating	Gender	
	Male	Female
	Satisfaction Rating	Satisfaction Rating
	SUM	SUM
Race		
White	13600.00	10152.00
Black	375.00	251.00

Output 6.8

CUSTTYPE Retail	Gender	
	Male	Female
	Satisfaction Rating	Satisfaction Rating
	SUM	SUM
Race		
White	6830.00	5203.00
Black	189.00	125.00

Fitting Tables to the Page

One way to save space with three-dimensional tables is to use the BOX=_PAGE_ option to reduce the space taken up by titles. But the biggest space-waster with three-dimensional tables is all of the page breaks. No matter how short your table, if you specify a page variable, then PROC TABULATE is going to put a page break between each table.

```
PROC TABULATE DATA=TEMP;
   CLASS EDUC RACE GENDER;
   VAR INCOME;
   TABLE EDUC, RACE, GENDER*INCOME*MEAN
      / BOX=_PAGE_;
RUN;
```

This code will produce three pages of tables. Each page will have a table that is only three rows high. The remaining part of each page is wasted.

Luckily, there is another TABLE statement option that can help us out. The CONDENSE option tells PROC TABULATE to fit more than one table on the page. If your tables are short enough that more than one will fit on the page, then PROC TABULATE will leave out the page breaks.

In the following code, you can see that the CONDENSE option is specified like the BOX= option. You put a slash at the end of the TABLE statement, and then list any of the options that you'd like to apply.

```
PROC TABULATE DATA=TEMP;
   CLASS EDUC RACE GENDER;
   VAR INCOME;
   TABLE EDUC, RACE, GENDER*INCOME*MEAN
      / BOX=_PAGE_ CONDENSE;
   TITLE "Tables With CONDENSE Option";
RUN;
```

In the result shown in Output 6.9, you can see that all three tables fit into a single page. The output in this book is shown exactly as it was created (no page breaks were removed). Each of these tables was short enough so that they could all fit on the page.

It is important to note that PROC TABULATE will not put part of the next table on the previous page. If there isn't room for the whole table, then PROC TABULATE will generate a page break and put the whole thing on the next page.

The only time PROC TABULATE will split a table between two pages is when even one table will not fit on a page. If the table is too wide or too tall to fit on a single page, then it will be split up. You will always see a "CONTINUED" label under the table when this happens.

Output 6.9

```
Tables With CONDENSE Option

---------------------------------------------------------------
|Education <HS                    |         Gender             |
|                                 |---------------------------|
|                                 |    Male     |   Female    |
|                                 |-----------+-------------|
|                                 |   Income    |   Income    |
|                                 |-----------+-------------|
|                                 |    MEAN     |    MEAN     |
|--------------------------------+-----------+-----------|
|Race                             |           |           |
|-------------------------------|           |           |
|White                            |   15450.66|     8925.37 |
|--------------------------------+-----------+-----------|
|Black                            |   10854.10|     7816.04 |
---------------------------------------------------------------


---------------------------------------------------------------
|Education HS                     |         Gender             |
|                                 |---------------------------|
|                                 |    Male     |   Female    |
|                                 |-----------+-------------|
|                                 |   Income    |   Income    |
|                                 |-----------+-------------|
|                                 |    MEAN     |    MEAN     |
|--------------------------------+-----------+-----------|
|Race                             |           |           |
|-------------------------------|           |           |
|White                            |   26627.88|    13715.56 |
|--------------------------------+-----------+-----------|
|Black                            |   18506.05|     8643.96 |
---------------------------------------------------------------


---------------------------------------------------------------
|Education College                |         Gender             |
|                                 |---------------------------|
|                                 |    Male     |   Female    |
|                                 |-----------+-------------|
|                                 |   Income    |   Income    |
|                                 |-----------+-------------|
|                                 |    MEAN     |    MEAN     |
|--------------------------------+-----------+-----------|
|Race                             |           |           |
|-------------------------------|           |           |
|White                            |   43899.85|    23681.98 |
|--------------------------------+-----------+-----------|
|Black                            |   33986.82|    20491.55 |
---------------------------------------------------------------
```

PART 3

Intermediate Topics

Handling Percentages

Why two chapters on a single statistic? Because percentages are different than other statistics. In most cases, changing the statistic displayed in a PROC TABULATE table is as simple as changing the statistic name in the TABLE statement.

For example, you can take your PROC TABULATE code and switch the statistic keyword MEAN to N or MAX with no other changes to the TABLE statement, and your new table will now show the number of observations or the maximum value in each cell instead of the mean.

Percentages are different. In all but the simplest cases, you need to change the syntax of the TABLE statement in order to get the percentage you want.

The Example

The goal in this chapter is to show how to build a table similar to the table below with cell percentages, row percentages, or column percentages. The examples will also show how to add row or column totals.

	Education			
	<HS	HS Graduate	College	
	PCTN	PCTN	PCTN	ALL
Occupation				
Managerial	30	20	50	100
Professional	10	40	50	100
Technical	20	40	40	100

If you want to learn how to build more complicated percentages, the following chapter discusses them in detail.

Starting with Cell Percentages

In this example, we are building a table that shows the relationship between occupation and education. Cell percentages can be a handy tool to help you build an informative table. A cell percentage tells you what percentage of the observations in the entire table is represented in each table cell. This is a good way to see if the data is distributed evenly between the categories of your row and column variables.

If too many observations are clustered in one cell of the table and other table cells are nearly empty, you may want to recalculate or reformat the row or column variables to achieve a more even distribution of observations. Otherwise, when you report a statistic like a mean for the observations in a table cell, readers of your table may draw inappropriate conclusions from cells with too little data to have any significance.

The best way to build a table involving percentages is to first build the table, and then change the statistic to a percentage. It is always better to work on the *structure* of a table first and the *content* of the table second. For this example, the structure is simple. The rows show a series of occupations. The columns show three categories of education. Notice that both the rows and columns are classification variables. The next example will show how to use percentages with continuous variables.

```
PROC TABULATE DATA=TEMP;
   CLASS OCCUP EDUC;
   TABLE OCCUP, EDUC;
   TITLE 'EXAMPLE BEFORE ADDING CELL PERCENTAGES';
RUN;
```

In Output 7.1, you can see that this TABLE statement produces a cross-tabulation of occupation and education, showing the number of observations in each cell. N is the default statistic for classification variables if no other statistic is specified. You can look at the values of N and get a rough sense of how the data is distributed in the table. In the cell labeled ❶, it appears as if an N of 130 represents a large number of observations, but it would be more helpful to know what percentage the cell represents of the total observations in the table. If there are 500 observations in the table, then 130 is a lot (26%). But if there are 50,000 observations in the table, then 130 is less than 0.3%.

What you need is a table of cell percentages. These are the easiest type of percentages to specify. The syntax is just like that of any other statistic. For cell percentages, you can use the statistic keyword PCTN just as you would use the statistic keyword N.

```
PROC TABULATE DATA=TEMP;
   CLASS OCCUP EDUC;
   TABLE OCCUP*PCTN, EDUC;
   TITLE 'EXAMPLE AFTER ADDING CELL PERCENTAGES';
RUN;
```

Now if you look at the cell labeled ❷ in Output 7.2, you can see that those 130 observations actually represent about 3% of the overall total. Looking at the cell percentages, you can see that this cross-tabulation appears to represent a fairly even distribution of the data. The only exception is the bottom row of the table (occupation "Farming"). Depending on the goals of your analysis, this row could be dropped or combined with another occupation.

Output 7.1

EXAMPLE BEFORE ADDING CELL PERCENTAGES

		Education		
		<HS	HS graduate	College
		N	N	N
Occupation				
Managerial		16.00	❶ 130.00	624.00
Professional		3.00	34.00	828.00
Technical		1.00	28.00	115.00
Sales		26.00	159.00	262.00
Clerical		16.00	301.00	359.00
Services		81.00	264.00	185.00
Manufacturing		140.00	381.00	212.00
Farming		18.00	31.00	30.00

Output 7.2

EXAMPLE AFTER ADDING CELL PERCENTAGES

		Education		
		<HS	HS graduate	College
Occupation				
Managerial	PCTN	0.38	❷ 3.06	14.70
Professional	PCTN	0.07	0.80	19.51
Technical	PCTN	0.02	0.66	2.71
Sales	PCTN	0.61	3.75	6.17
Clerical	PCTN	0.38	7.09	8.46
Services	PCTN	1.91	6.22	4.36
Manufacturing	PCTN	3.30	8.98	5.00
Farming	PCTN	0.42	0.73	0.71

Comparing PCTN and PCTSUM

If you are interested in seeing how the observations in a data set are distributed between classification variables, then PCTN is the statistic for you. But what if you want to see how the total *value* of an analysis variable is distributed between the cells of a table?

For example, if you're building a table showing mean income by occupation and education, you might want to know how the total sum of income in the data set is distributed between the values of occupation and education.

Instead of looking at the percentage of the total *number of observations* that is in each cell, you want to know the percentage of the total *sum* that is in each cell. So instead of using the PCTN statistic, we will use the PCTSUM statistic.

Just as in the previous example, we start by building the table structure. The table has two classification variables (OCCUP and EDUC) and one analysis variable (INCOME).

```
PROC TABULATE DATA=TEMP;
   CLASS OCCUP EDUC;
   VAR INCOME;
   TABLE OCCUP*INCOME, EDUC;
RUN;
```

This table is shown in Output 7.3. Because no statistic was specified, and there is an analysis variable in the table cells, PROC TABULATE defaults to the SUM statistic.

It's no great surprise that most of the income is earned by college graduates, rather than by people who do not have a college degree. It also appears that managers and professionals represent a large share of the total income. To quantify this difference, we can compute the percent sum statistic.

For cell percentages, the syntax for PCTSUM is just like the syntax for the PCTN. You just add the PCTSUM statistic to the row or column dimension of the TABLE statement. The only difference between the syntax for PCTN and PCTSUM is that PCTSUM requires an analysis variable, but PCTN only needs a classification variable.

```
PROC TABULATE DATA=TEMP;
   CLASS OCCUP EDUC;
   VAR INCOME;
   TABLE OCCUP*INCOME*PCTSUM, EDUC;
RUN;
```

Now you can see in Output 7.4 that the college graduates, especially the managers and professionals, grab a disproportionate share of the total income pool. Together, these two table cells represent over 45% of the income in this population.

Output 7.3

			<HS	HS graduate	College
Occupation					
Managerial	Income	SUM	402354.00	4342657.00	30488451.00
Professional	Income	SUM	55823.00	986612.00	33958471.00
Technical	Income	SUM	19615.00	865932.00	3429943.00
Sales	Income	SUM	485683.00	3723540.00	10181363.00
Clerical	Income	SUM	294999.00	6508427.00	8448656.00
Services	Income	SUM	1076693.00	4390752.00	3985256.00
Manufacturing	Income	SUM	2875597.00	10299079.00	6488467.00
Farming	Income	SUM	301448.00	574994.00	546543.00

Output 7.4

			<HS	HS graduate	College
Occupation					
Managerial	Income	PCTSUM	0.30	3.22	22.63
Professional	Income	PCTSUM	0.04	0.73	25.20
Technical	Income	PCTSUM	0.01	0.64	2.55
Sales	Income	PCTSUM	0.36	2.76	7.56
Clerical	Income	PCTSUM	0.22	4.83	6.27
Services	Income	PCTSUM	0.80	3.26	2.96
Manufacturing	Income	PCTSUM	2.13	7.64	4.82
Farming	Income	PCTSUM	0.22	0.43	0.41

Calculating Row and Column Percentages

Okay, so now you're wondering what's so hard about PROC TABULATE percentages? So far, PCTN and PCTSUM seem like any other statistic. But the previous examples showed how to create simple tables with *cell* percentages; *row* and *column* percentages are another matter altogether.

PROC TABULATE gives you the power to define whether you want a row or column percentage when you use the PCTN or PCTSUM statistic. How do you do this? Unfortunately, it's not as simple as specifying ROW or COLUMN. Instead, you must specify a denominator. The syntax for specifying a denominator is as follows:

> **PCTN<denominator>** or **PCTSUM<denominator>**

Between the two angle brackets, you list the name of the variable that will be used as the denominator in calculating the percentages. To calculate the PCTN statistic, PROC TABULATE divides the number of observations in each cell by this denominator to come up with a row or column percent. If you don't specify a denominator, PROC TABULATE uses the total number of observations in the table as the denominator, and you end up with cell percentages (as in the previous examples).

In this example, the goal is to calculate both row and column percentages. For the row percentages, you need to use the total number of observations in the row as the denominator. To get the number of observations in the row, you add up the observations for each column in that row. So to get the row percentages, you need to specify the column variable as the denominator. In this example, we need to specify education as the denominator.

```
PROC TABULATE DATA=TEMP;
   CLASS OCCUP EDUC;
   TABLE OCCUP*PCTN<EDUC>, EDUC;
   TITLE 'EXAMPLE SHOWING ROW PERCENTAGES';
RUN;
```

To see how PROC TABULATE works, look at the cell labeled ❶ in Output 7.5. For this cell, PROC TABULATE takes the numerator (the number of observations with education <HS and occupation MANAGERIAL) and divides it by the denominator (the number of observations with occupation MANAGERIAL for all values of education) to compute the row percentage.

To compute column percentages, switch the denominator from the column variable to the row variable. It does not matter whether the PCTN statistic is listed in the row or the column dimension of the TABLE statement.

```
PROC TABULATE DATA=TEMP;
   CLASS OCCUP EDUC;
   TABLE OCCUP*PCTN<OCCUP>, EDUC;
   TITLE 'EXAMPLE SHOWING COLUMN PERCENTAGES';
RUN;
```

To see the difference between the two tables, look at the cell labeled ❷ in Output 7.6. For this cell, PROC TABULATE takes the numerator (the number of observations with education <HS and occupation MANAGERIAL) and divides it by the denominator (the number of observations with education <HS for all values of occupation) to compute the row percentage.

Output 7.5

EXAMPLE SHOWING ROW PERCENTAGES

		Education		
		<HS	HS graduate	College
Occupation				
Managerial	PCTN	2.08	❶ 16.88	81.04
Professional	PCTN	0.35	3.93	95.72
Technical	PCTN	0.69	19.44	79.86
Sales	PCTN	5.82	35.57	58.61
Clerical	PCTN	2.37	44.53	53.11
Services	PCTN	15.28	49.81	34.91
Manufacturing	PCTN	19.10	51.98	28.92
Farming	PCTN	22.78	39.24	37.97

Output 7.6

EXAMPLE SHOWING COLUMN PERCENTAGES

		Education		
		<HS	HS graduate	College
Occupation				
Managerial	PCTN	5.32	❷ 9.79	23.86
Professional	PCTN	1.00	2.56	31.66
Technical	PCTN	0.33	2.11	4.40
Sales	PCTN	8.64	11.97	10.02
Clerical	PCTN	5.32	22.67	13.73
Services	PCTN	26.91	19.88	7.07
Manufacturing	PCTN	46.51	28.69	8.11
Farming	PCTN	5.98	2.33	1.15

Don't worry, you don't have to understand how denominators work to use PROC TABULATE to build a simple two-variable table. Follow this rule: If you want the percentages in each *column* to add up to 100%, select the *row* variable as the denominator. If you want the percentages in each *row* to add up to 100%, select the *column* variable as the denominator.

Adding Row Totals

The next two examples show how to add totals to your tables. This example shows row totals, and the following example is the same table with column totals added. If you want to add totals to a table with row percentages, stay on this page. Otherwise, turn the page and look at the column percentage example.

When you build a table with row percentages, the best way to clearly identify what kind of percentages you are displaying is to include row totals. This way, the end user of your table can see that the percentages in each row add up to a total of 100%.

Just as in the sections in Chapter 5 titled "Adding Totals to the Rows and Columns" and "Adding Subtotals," we need to use the ALL keyword to add a total to the table. As a first step, we build a table that shows the correct structure, but without the percentages.

```
PROC TABULATE DATA=TEMP;
   CLASS OCCUP EDUC;
   TABLE OCCUP, EDUC ALL;
RUN;
```

This table is shown in Output 7.7. Now that the basic table has been constructed, the next step is to add the percentages. Because we are adding row percentages, we use the column dimension as the denominator. What makes this table more challenging is that instead of having education as a column header, we now have the column total ALL.

For PROC TABULATE to correctly compute the denominator, you have to use the entire column definition in the denominator definition. If there were no column total involved, the TABLE statement would be.

```
TABLE OCCUP, EDUC*PCTN<EDUC>;
```

But because this table has a more complex column definition, we use that same definition for the PCTN denominator:

```
TABLE OCCUP, (EDUC ALL)*PCTN<EDUC ALL>;
```

The parentheses are added around the column definition so that the PCTN definition will apply to all columns in the table. So the final PROC TABULATE code for row percentages with totals is as follows:

```
PROC TABULATE DATA=TEMP;
   CLASS OCCUP EDUC;
   TABLE OCCUP, (EDUC ALL)*PCTN<EDUC ALL>;
RUN;
```

Now all it takes is a quick look at the table in Output 7.8 to see that the percentages displayed are row percentages and that they add up to 100% for each row.

Output 7.7

	Education			
	<HS	HS graduate	College	ALL
	N	N	N	N
Occupation				
Managerial	16.00	130.00	624.00	770.00
Professional	3.00	34.00	828.00	865.00
Technical	1.00	28.00	115.00	144.00
Sales	26.00	159.00	262.00	447.00
Clerical	16.00	301.00	359.00	676.00
Services	81.00	264.00	185.00	530.00
Manufacturing	140.00	381.00	212.00	733.00
Farming	18.00	31.00	30.00	79.00

Output 7.8

	Education			
	<HS	HS graduate	College	ALL
	PCTN	PCTN	PCTN	PCTN
Occupation				
Managerial	2.08	16.88	81.04	100.00
Professional	0.35	3.93	95.72	100.00
Technical	0.69	19.44	79.86	100.00
Sales	5.82	35.57	58.61	100.00
Clerical	2.37	44.53	53.11	100.00
Services	15.28	49.81	34.91	100.00
Manufacturing	19.10	51.98	28.92	100.00
Farming	22.78	39.24	37.97	100.00

Adding Column Totals

This example is the same table as the previous example with column totals added. If you want to add totals to a table with row percentages, turn back one page to see the row percentage example.

When you build a table with column percentages, the best way to clearly identify what kind of percentages you are displaying is to include column totals. This way, the end user of your table can see that the percentages in each column add up to a total of 100%.

Just as in the Sections in Chapter 5 titled "Adding Totals to the Rows and Columns" and "Adding Subtotals," we need to use the ALL keyword to add a total to the table. As a first step, we build a table that shows the correct structure, but without the percentages.

```
PROC TABULATE DATA=TEMP;
   CLASS OCCUP EDUC;
   TABLE OCCUP ALL, EDUC;
RUN;
```

This output is shown in Output 7.9. Now that the basic table has been constructed, the next step is to add the percentages. Because we are adding column percentages, we use the row dimension as the denominator. What makes this table more challenging is that instead of having education as a row header, we now have the row total ALL.

For PROC TABULATE to correctly compute the denominator, you have to use the entire row definition in the denominator definition. If there was no row total involved, the TABLE statement would be as follows:

```
TABLE OCCUP, EDUC*PCTN<OCCUP>;
```

But because this table has a more complex row definition, we use that same definition for the PCTN denominator:

```
TABLE OCCUP ALL, EDUC*PCTN<OCCUP ALL>;
```

So the final PROC TABULATE code for column percentages with totals is as follows:

```
PROC TABULATE DATA=TEMP;
   CLASS OCCUP EDUC;
   TABLE OCCUP ALL, EDUC*PCTN<OCCUP ALL>;
RUN;
```

Now all it takes is a quick look at the table in Output 7.10 to see that the percentages displayed are column percentages and that they add up to 100% for each column.

Actually, when you build tables with totals, you will notice that the percentages do not always add up to the displayed total of 100%. If you take out a calculator, you will notice that in some tables, some columns add up to 99.9% or 100.1% due to rounding. You can always add a footnote to your tables to explain this.

Output 7.9

	Education		
	<HS	HS graduate	College
	N	N	N
Occupation			
Managerial	16.00	130.00	624.00
Professional	3.00	34.00	828.00
Technical	1.00	28.00	115.00
Sales	26.00	159.00	262.00
Clerical	16.00	301.00	359.00
Services	81.00	264.00	185.00
Manufacturing	140.00	381.00	212.00
Farming	18.00	31.00	30.00
ALL	301.00	1328.00	2615.00

Output 7.10

	Education		
	<HS	HS graduate	College
	PCTN	PCTN	PCTN
Occupation			
Managerial	5.32	9.79	23.86
Professional	1.00	2.56	31.66
Technical	0.33	2.11	4.40
Sales	8.64	11.97	10.02
Clerical	5.32	22.67	13.73
Services	26.91	19.88	7.07
Manufacturing	46.51	28.69	8.11
Farming	5.98	2.33	1.15
ALL	100.00	100.00	100.00

Handling Percentages with Complex Denominators

So far, so good. We've built a series of basic tables with percentages. We have built cell percentages, row percentages, and column percentages. But all of the tables we have seen so far have been relatively simple. Percentages get a lot more challenging when the table becomes more complex.

This chapter shows you how to build complex tables with complex percentages. These tables don't have to be intimidating if you know how to break them down and build them step by step.

The Example

This chapter shows how to build a variety of complex tables, including the following table with nested row variables, column percentages, subtotals, and totals.

| | | Customer Type | |
| | | Retail | Wholesale |
		PCTN	PCTN
Race	Gender		
White	Female	30	15
	Male	10	15
	ALL	40	30
Black	Gender		
	Female	20	40
	Male	40	30
	ALL	60	70
ALL		100	100

Percentages for Multiple Row Variables

Our first table shows row percentages. What's different about this table is that this time there are multiple row variables. So we'll be building percentages for more than one variable at a time.

Our starting point is a table showing a breakdown of customers by race, gender, business type, and business size. The following code produces a table that shows the number of customers in each category:

```
PROC TABULATE DATA=TEMP;
   CLASS GENDER RACE CUSTTYPE SIZE;
   TABLE (RACE CUSTTYPE)*GENDER, SIZE*N;
RUN;
```

But it would be much more interesting to see some percentages. This will make it easier to see the distribution of customers between the categories.

```
PROC TABULATE DATA=TEMP;
   CLASS GENDER RACE CUSTTYPE SIZE;
   TABLE (RACE CUSTTYPE)*GENDER, SIZE*PCTN;
RUN;
```

But if all you do is switch the N for a PCTN, you get the table shown in Output 8.1. Because no denominator was specified, PROC TABULATE guessed at what denominator to use. Every once in a while, you get lucky and this guess is correct, but more often you get a nonsense table like Output 8.1. What we got was cell percentages for each combination of RACE*GENDER*SIZE and CUSTTYPE*GENDER*SIZE.

PROC TABULATE came up with the definition RACE*GENDER*SIZE CUSTTYPE*GENDER*SIZE by expanding out the crossings of the table. This is done by crossing the row definition with the column definition, and then expanding out all of the items in parentheses.

    (RACE CUSTTYPE)*GENDER*SIZE   *row definition multiplied by column dimension*

    RACE*GENDER*SIZE CUSTTYPE*GENDER*SIZE   *parentheses removed*

Whenever you have trouble figuring out what is going on in a complex table with percentages, take the time to do this expansion of the table definition and see what's going on. Then, to get a different definition, all you have to do is pick the part you want. To be a legal denominator, you have to include at least one variable from each part of the expanded table definitions. For example, let's modify the code to get a breakdown by gender for each category. GENDER is a legal denominator definition because this variable is present in each part of the expanded definition. We can pick this as our denominator definition, as in the following example:

```
TABLE (RACE CUSTTYPE)*GENDER, SIZE*PCTN<GENDER>;
```

The output is shown in Output 8.2. It's a little hard to see, but if you add up the percentages for male and female for each combination of race and size or customer type and size, you'll see that they add up to 100%. This table would be clearer if we added totals at the bottom of each grouping. To do that, we would use the following code (output not shown):

```
TABLE (RACE CUSTTYPE)*(GENDER ALL), SIZE*PCTN<GENDER ALL>;
```

Output 8.1

		Size	
		Large	Small
		PCTN	PCTN
Race	Gender		
White	Male	17.63	26.20
	Female	23.21	30.34
Black	Male	0.44	0.76
	Female	0.53	0.91
Customer Type			
Retail	Male	9.07	13.62
	Female	11.85	16.22
Wholesale	Male	8.99	13.34
	Female	11.89	15.02

Output 8.2

		Size	
		Large	Small
		PCTN	PCTN
Race	Gender		
White	Male	43.17	46.34
	Female	56.83	53.66
Black	Male	45.00	45.63
	Female	55.00	54.37
Customer Type			
Retail	Male	43.35	45.64
	Female	56.65	54.36
Wholesale	Male	43.07	47.04
	Female	56.93	52.96

Percentages for Multiple Column Variables

This next example looks at a similar table, but this time we are computing percentages for multiple column variables. Again, we'll be building percentages for more than one variable at a time.

Our starting point is a table similar to the one in the previous examples. It shows a breakdown of customers by race, gender, business type, and business size. Only this time, there are multiple row variables and multiple column variables. The following code produces a table that shows the number of customers in each category.

```
PROC TABULATE DATA=TEMP;
   CLASS GENDER RACE CUSTTYPE SIZE;
   TABLE RACE*GENDER, (CUSTTYPE SIZE)*N;
RUN;
```

Again, it would be much more interesting to see some percentages. This makes it easier to see the distribution of customers between the categories.

```
PROC TABULATE DATA=TEMP;
   CLASS GENDER RACE CUSTTYPE SIZE;
   TABLE RACE*GENDER, (CUSTTYPE SIZE)*PCTN;
RUN;
```

But if all you do is switch the N for a PCTN, you get the table shown in Output 8.3. Because no denominator was specified, PROC TABULATE used cell percentages. You can figure out this default definition by expanding out the crossings of the table, as explained in the previous example. For this table, you get the following expansion:

RACE*GENDER*CUSTTYPE RACE*GENDER*SIZE

However, these cell percentages are not useful. What would be more helpful is to see the percentage breakdown between categories of each of the two column variables. In other words, we want row percentages. To build row percentages, we use the column dimension as the denominator. The following definition tells PROC TABULATE to create percentages that add up to 100% when you sum the values of each row for CUSTTYPE, and then percentages that add up to 100% when you sum the values of each row for SIZE:

TABLE RACE*GENDER, (CUSTTYPE SIZE)*PCTN<CUSTTYPE SIZE>;

The output is shown in Output 8.4. It's a little hard to see, but if you add up the percentages for retail and wholesale for each combination of race and gender, you'll see that they add up to 100%. You will also get 100% if you add up the percentages for large and small for each combination of race and gender.

This table would be clearer if we added totals at the right of each grouping. To do that, use the following code:

TABLE RACE*GENDER, (CUSTTYPE ALL SIZE ALL)*PCTN<CUSTTYPE ALL SIZE ALL>;

The resulting table is shown in Output 8.5. The only change made in this code is to add two totals columns to the column definition, and then to revise the PCTN definition to match. Actually, the two totals columns are identical. The values are repeated in this example to aid the reader in interpreting the table.

Output 8.3

		Customer Type		Size	
		Retail	Wholesale	Large	Small
		PCTN	PCTN	PCTN	PCTN
Race	Gender				
White	Male	22.04	21.78	17.63	26.20
	Female	27.38	26.16	23.21	30.34
Black	Male	0.65	0.55	0.44	0.76
	Female	0.70	0.74	0.53	0.91

Output 8.4

		Customer Type		Size	
		Retail	Wholesale	Large	Small
		PCTN	PCTN	PCTN	PCTN
Race	Gender				
White	Male	50.30	49.70	40.22	59.78
	Female	51.13	48.87	43.34	56.66
Black	Male	54.05	45.95	36.49	63.51
	Female	48.31	51.69	37.08	62.92

Output 8.5

		Customer Type			Size		
		Retail	Wholesale	ALL	Large	Small	ALL
		PCTN	PCTN	PCTN	PCTN	PCTN	PCTN
Race	Gender						
White	Male	50.30	49.70	100.00	40.22	59.78	100.00
	Female	51.13	48.87	100.00	43.34	56.66	100.00
Black	Male	54.05	45.95	100.00	36.49	63.51	100.00
	Female	48.31	51.69	100.00	37.08	62.92	100.00

Picking the Right Denominator

The most confusing thing about complex tables with percentages is that there is no single correct denominator. There are many ways that you can set up the denominator. This example shows how to take a single TABLE statement and build a variety of different tables by swapping the denominator. Our starting point is the table from the previous example:

```
PROC TABULATE DATA=TEMP;
   CLASS GENDER RACE CUSTTYPE SIZE;
   TABLE RACE*GENDER, (CUSTTYPE SIZE)*PCTN;
RUN;
```

As we saw before, this code will generate cell percentages. The actual denominator definition is

```
RACE*GENDER*CUSTTYPE RACE*GENDER*SIZE
```

But there are a number of other definitions that will work. There are also a number of definitions that won't work, as detailed in the following table:

Denominator	What you get	Results
RACE*GENDER	column percentages	Output 8.6
CUSTTYPE SIZE	row percentages for each column variable	Output 8.4
GENDER	column percentages of each gender for each race	Output 8.7
RACE	column percentages of each race for each gender (very hard to read–if you want these percentages, you should change the row definition from RACE*GENDER to GENDER*RACE first).	not shown
CUSTTYPE	ERROR: PROC TABULATE can't figure out what to do with this definition because it doesn't make sense for the two columns showing SIZE. Denominator definitions have to make sense for all cells in the table.	not shown
SIZE	ERROR: PROC TABULATE can't figure out what to do with this definition either.	not shown
RACE*CUSTTYPE RACE*SIZE	1. Percentages that add to 100% for all females and customer types for each race (repeats for males). 2. Percentages that add to 100% for all females and sizes for each race (repeats for males).	Output 8.8
GENDER* CUSTTYPE GENDER*SIZE	1. Percentages that add to 100% for all whites and all customer types (repeats for blacks). 2. Percentages that add to 100% for all whites and all sizes (repeats for blacks).	not shown

The best way to understand how all of these definitions work is to build a sample table and start experimenting with the denominators. Just like the examples in the table above, you will find some that work, some that work but make no sense, and some that do not work at all.

Output 8.6

		Customer Type		Size	
		Retail	Wholesale	Large	Small
		PCTN	PCTN	PCTN	PCTN
Race	Gender				
White	Male	43.42	44.24	42.17	45.01
	Female	53.93	53.14	55.51	52.13
Black	Male	1.27	1.12	1.04	1.31
	Female	1.37	1.51	1.28	1.56

Output 8.7

		Customer Type		Size	
		Retail	Wholesale	Large	Small
		PCTN	PCTN	PCTN	PCTN
Race	Gender				
White	Male	44.60	45.43	43.17	46.34
	Female	55.40	54.57	56.83	53.66
Black	Male	48.19	42.50	45.00	45.63
	Female	51.81	57.50	55.00	54.37

Output 8.8

		Customer Type		Size	
		Retail	Wholesale	Large	Small
		PCTN	PCTN	PCTN	PCTN
Race	Gender				
White	Male	48.96	48.38	39.15	58.19
	Female	49.79	47.59	42.21	55.18
Black	Male	1.44	1.22	0.97	1.69
	Female	1.26	1.35	0.97	1.65

Percentages Mixed with Other Statistics

Each of the previous examples has shown a table with percentages alone, but there's no requirement that percentages must be shown alone. It's perfectly legal to add other statistics to the same table.

For example, you can take a table that shows column percentages and add a column that shows the mean of some other variable.

```
PROC TABULATE DATA=TEMP;
   CLASS GENDER RACE SAT;
   VAR INCOME;
   TABLE GENDER RACE,
      SAT*PCTN<GENDER RACE> INCOME*MEAN;
RUN;
```

This produces the table shown in Output 8.9. You can add as many new variables and statistics to the column dimension as you like. One thing you can't do is the following TABLE statement:

```
TABLE GENDER RACE,
   SAT*PCTN<GENDER RACE> SAT*MEAN;
```

While it would be nice to follow the percentages with a mean satisfaction score, it can't be done with this code. This is because percentages require a CLASS variable and MEAN requires an analysis variable. SAT can't be both. For a way to get around this, see Chapter 15, "PROC TABULATE Tricks: How to Cheat to Create Complex Tables."

There is another way to show a percentage and a MEAN that use the same variable, and that's to do a PCTSUM instead of a PCTN. This allows percentages with information about the analysis variable used for the means.

```
PROC TABULATE DATA=TEMP;
   CLASS GENDER RACE SAT;
   VAR INCOME;
   TABLE GENDER, SAT*PCTSUM<GENDER>*INCOME MEAN*INCOME;
RUN;
```

This produces the output shown in Output 8.10. Now the percentages and MEAN have something in common. They all involve the variable INCOME.

This final example shows how you can add another statistic to a percentages table by varying the statistic selected for each row. In this example, one row variable is shown as a percentage and one is shown as an N.

```
PROC TABULATE DATA=TEMP;
   CLASS GENDER RACE SAT;
   TABLE GENDER*PCTN<SAT> RACE*N, SAT;
RUN;
```

This table, shown in Output 8.11, works because both PCTN and N are statistics that work for CLASS variables. PROC TABULATE can show the percentage of observations for each value of SAT and the N for each value of SAT in the same table.

Output 8.9

		Satisfaction Rating					
		1	2	3	4	5	Income
		PCTN	PCTN	PCTN	PCTN	PCTN	MEAN
Gender							
Male		47.44	45.93	45.63	42.63	43.64	34087.69
Female		52.56	54.07	54.37	57.37	56.36	17902.92
Race							
White		97.23	97.12	96.48	98.20	97.75	25344.29
Black		2.77	2.88	3.52	1.80	2.25	19460.60

Output 8.10

		Satisfaction Rating					
		1	2	3	4	5	
		PCTSUM	PCTSUM	PCTSUM	PCTSUM	PCTSUM	MEAN
		Income	Income	Income	Income	Income	Income
Gender							
Male		64.23	61.84	60.07	58.36	60.04	34087.69
Female		35.77	38.16	39.93	41.64	39.96	17902.92

Output 8.11

		Satisfaction Rating				
		1	2	3	4	5
Gender						
Male	PCTN	20.29	20.65	20.04	19.54	19.47
Female	PCTN	18.41	19.91	19.56	21.53	20.59
Race						
White	N	1158.00	1216.00	1180.00	1253.00	1214.00
Black	N	33.00	36.00	43.00	23.00	28.00

Percentages with Subtotals

After you start building complex tables with percentages, overall totals may not be enough to make the tables understandable for the reader. You can also add subtotals to make things clearer. With subtotals, even an extremely long table can be easily understood.

For example, when you have a table with nested classifications, you may want to show how things add up within the nestings as well as overall. To illustrate the use of subtotals, we can use the basic table that we've been working with all along in this chapter. This code produces a basic table with column percentages.

```
PROC TABULATE DATA=TEMP;
   CLASS GENDER RACE CUSTTYPE;
   TABLE RACE*GENDER ALL,
      CUSTTYPE*PCTN<RACE*GENDER ALL>;
RUN;
```

As you can see from Output 8.12, this table has totals that show that each column totals to 100%. But wouldn't it be nice to have subtotals for each race? To do that, we need to add these subtotals to the row dimension. This is done by adding the ALL variable to the row dimension. By putting it in parentheses after GENDER, we will get a subtotal for each race following the breakdown by gender.

```
TABLE RACE*(GENDER ALL) ALL,
   CUSTTYPE*PCTN< denominator goes here >;
```

This is probably the hardest part of the example. Once the rows and columns are set up correctly, creating the right denominator definition is relatively easy. You may want to test your code at this stage without the percentages to see if you've got the totals added correctly.

Now we need to fix the denominator definition. You might think that all we have to do to the PCTN definition is revise it the same way we revised the row definition.

```
TABLE RACE*(GENDER ALL) ALL,
   CUSTTYPE*PCTN<RACE*(GENDER ALL) ALL>;
```

Well, you were almost right. If you run the following code, you would get the error message shown in Output 8.13. Unfortunately, PROC TABULATE can't figure out parentheses inside the denominator definition. To get this code to work, you have to multiply out RACE*(GENDER ALL) and use RACE*GENDER RACE*ALL as in the following code:

```
PROC TABULATE DATA=TEMP;
   CLASS GENDER RACE CUSTTYPE;
   TABLE RACE*(GENDER ALL) ALL,
      CUSTTYPE*PCTN<RACE*GENDER RACE*ALL ALL>;
RUN;
```

Now we get the table shown in Output 8.14. It has subtotals after each race, which makes it even easier to see that this data set has very little data for RACE=Black.

Output 8.12

		Customer Type	

		Retail	Wholesale
		-------------+-----------
		PCTN	PCTN
----------------------------+-------------+-----------
Race	Gender		
--------------+-------------		
White	Male	43.42	44.24
	-------------+-------------+-----------
	Female	53.93	53.14
--------------+-------------+-------------+-----------
Black	Male	1.27	1.12
	-------------+-------------+-----------
	Female	1.37	1.51
--------------+-------------+-------------+-----------
ALL		100.00	100.00

Output 8.13

```
6    PROC TABULATE DATA=TEMP;
7         CLASS GENDER RACE CUSTTYPE SIZE;
8         TABLE RACE*(GENDER ALL) ALL, CUSTTYPE*PCTN<RACE*(GENDER ALL) ALL>;
                                                    -
                                                    22
9    RUN;

ERROR 22-322: Expecting one of the following: { NAME > }.  The statement is being ignored.
```

Output 8.14

		Customer Type	

		Retail	Wholesale
		-------------+-----------
		PCTN	PCTN
----------------------------+-------------+-----------
Race	Gender		
--------------+-------------		
White	Male	43.42	44.24
	-------------+-------------+-----------
	Female	53.93	53.14
	-------------+-------------+-----------
	ALL	97.36	97.37
--------------+-------------+-------------+-----------
Black	Gender		

	Male	1.27	1.12
	-------------+-------------+-----------
	Female	1.37	1.51
	-------------+-------------+-----------
	ALL	2.64	2.63
--------------+-------------+-------------+-----------
ALL		100.00	100.00

Percentage Subtotals on Both Dimensions

This example is included more as an academic exercise than as something you'd want to use in the real world. If you end up building a table this complicated and you plan to add percentages, then you probably want to rethink your table design. However, the table is a good example of how to figure out complex denominator definitions.

The following code produces our sample table. It shows race broken down by gender versus customer type broken down by size. There are subtotals and totals along both dimensions. The code below creates the basic table without percentages, so that you can see how it works.

```
PROC TABULATE DATA=TEMP;
    CLASS GENDER RACE CUSTTYPE SIZE;
    TABLE RACE*(GENDER ALL) ALL, (CUSTTYPE*(SIZE ALL) ALL)*N;
RUN;
```

The resulting table is shown in Output 8.15. This is an extremely complex table, but it is possible to add percentages. First, we have a choice: do we want to add row percentages, column percentages, or cell percentages?

To get row percentages, we use the column dimension as our denominator. The resulting TABLE statement is as follows:

```
TABLE RACE*(GENDER ALL) ALL, (CUSTTYPE*(SIZE ALL) ALL)*
    PCTN<CUSTTYPE*SIZE CUSTTYPE*ALL ALL>;
```

This definition was created by multiplying all of the items in parentheses within the column definition. Using this TABLE statement produces the table shown in Output 8.16. To get column percentages, the process is similar. If you multiply out the row dimension, you get the following TABLE statement (the output is not shown):

```
TABLE RACE*(GENDER ALL) ALL, (CUSTTYPE*(SIZE ALL) ALL)*
    PCTN<RACE*GENDER RACE*ALL ALL>;
```

The hardest type of percentage to set up is cell percentages. To get the definition, you have to expand the row and column dimensions. The following lines show how this is done, step by step.

```
(RACE*(GENDER ALL) ALL) * (CUSTTYPE*(SIZE ALL) ALL)

(RACE*GENDER RACE*ALL ALL) * (CUSTTYPE*SIZE CUSTTYPE*ALL ALL)

RACE*GENDER*CUSTTYPE*SIZE RACE*GENDER*CUSTTYPE*ALL RACE*GENDER*ALL
    RACE*ALL*CUSTTYPE*SIZE RACE*ALL*CUSTTYPE*ALL RACE*ALL*ALL
    ALL*CUSTTYPE*SIZE ALL*CUSTTYPE*ALL ALL*ALL
```

Yikes! This is a ridiculously complex denominator definition. Luckily, if you want cell percentages, you don't have to figure this out. Remember, cell percentages are the default for PROC TABULATE. So the TABLE statement for cell percentages is actually quite simple:

```
TABLE RACE*(GENDER ALL) ALL, (CUSTTYPE*(SIZE ALL) ALL)*PCTN;
```

Output 8.15

		Customer Type						
		Retail			Wholesale			
		Size			Size			
		Large	Small	ALL	Large	Small	ALL	ALL
		N	N	N	N	N	N	N
Race	Gender							
White	Male	547.00	816.00	1363.00	543.00	804.00	1347.00	2710.00
	Female	715.00	978.00	1693.00	720.00	898.00	1618.00	3311.00
	ALL	1262.00	1794.00	3056.00	1263.00	1702.00	2965.00	6021.00
Black	Gender							
	Male	14.00	26.00	40.00	13.00	21.00	34.00	74.00
	Female	18.00	25.00	43.00	15.00	31.00	46.00	89.00
	ALL	32.00	51.00	83.00	28.00	52.00	80.00	163.00
ALL		1294.00	1845.00	3139.00	1291.00	1754.00	3045.00	6184.00

Output 8.16

		Customer Type						
		Retail			Wholesale			
		Size			Size			
		Large	Small	ALL	Large	Small	ALL	ALL
		PCTN	PCTN	PCTN	PCTN	PCTN	PCTN	PCTN
Race	Gender							
White	Male	20.18	30.11	50.30	20.04	29.67	49.70	100.00
	Female	21.59	29.54	51.13	21.75	27.12	48.87	100.00
	ALL	20.96	29.80	50.76	20.98	28.27	49.24	100.00
Black	Gender							
	Male	18.92	35.14	54.05	17.57	28.38	45.95	100.00
	Female	20.22	28.09	48.31	16.85	34.83	51.69	100.00
	ALL	19.63	31.29	50.92	17.18	31.90	49.08	100.00
ALL		20.92	29.84	50.76	20.88	28.36	49.24	100.00

Percentages for Three-Dimensional Tables

It is possible to add percentages to three-dimensional tables. It's not difficult to write the code to create three-dimensional percentages. The challenge is in understanding what the resulting tables are showing. Percentages can be rather confusing across three-dimensional tables because you can't always see how they add up when looking at a single page.

For our example, we'll use the following table:

```
PROC TABULATE DATA=TEMP;
   CLASS SAT RACE GENDER;
   TABLE SAT ALL, RACE ALL, (GENDER ALL)*N;
RUN;
```

This produces a table with a page for each level of satisfaction plus a totals page. Each page shows race by gender with totals for the rows and columns. For this example, we walk through how to modify the TABLE statement to produce four kinds of percentages: row percentages, column percentages, page percentages, and cell percentages.

To build row percentages, you modify the TABLE statement just as you would if this was a two-dimensional table. You take the column definition as your denominator.

```
TABLE SAT ALL, RACE ALL, (GENDER ALL)*PCTN<GENDER ALL>;
```

To build column percentages, again you treat this like a two-dimensional table. You take the row definition as your denominator.

```
TABLE SAT ALL, RACE ALL, (GENDER ALL)*PCTN<RACE ALL>;
```

To get page percentages (these are cell percentages for each page), you follow a similar rule. Just as you take the row definition to get column percentages, and the column definition to get row percentages, you take the row definition crossed with the column dimension to get page percentages.

```
TABLE SAT ALL, RACE ALL, (GENDER ALL)*
      PCTN<RACE*GENDER RACE*ALL ALL*GENDER ALL*ALL>;
```

The resulting tables are shown in Output 8.17. Notice how the lower right corner of each table shows a total of 100%, so it is easy to see how the percentages on each page add up.

There is one other way that you can do percentages for a three-dimensional table: cell percentages. This means that if you add up all of the cells across all of the tables, you get 100%. Even if you put subtotals and totals on each page, the only place where they will add up to 100% is on the final page (assuming you create a totals page). This is an extremely confusing table and is not recommended. The code is presented here. Try it out if you're curious.

```
TABLE SAT ALL, RACE ALL, (GENDER ALL)*PCTN;
```

Output 8.17

Satisfaction Rating POOR-GOOD

```
--------------------------------------------------------------------
|                              |            Gender         |        |
|                              |---------------------------|        |
|                              |   Male   |  Female  |   ALL   |
|                              |----------+----------+----------|
|                              |   PCTN   |   PCTN   |   PCTN   |
|------------------------------+----------+----------+----------|
|Race                          |          |          |          |
|------------------------------|          |          |          |
|White                         |    44.98 |    51.96 |    96.94 |
|------------------------------+----------+----------+----------|
|Black                         |     1.34 |     1.72 |     3.06 |
|------------------------------+----------+----------+----------|
|ALL                           |    46.32 |    53.68 |   100.00 |
--------------------------------------------------------------------
```

Satisfaction Rating VERY GOOD-EXCELLENT

```
--------------------------------------------------------------------
|                              |            Gender         |        |
|                              |---------------------------|        |
|                              |   Male   |  Female  |   ALL   |
|                              |----------+----------+----------|
|                              |   PCTN   |   PCTN   |   PCTN   |
|------------------------------+----------+----------+----------|
|Race                          |          |          |          |
|------------------------------|          |          |          |
|White                         |    42.14 |    55.84 |    97.97 |
|------------------------------+----------+----------+----------|
|Black                         |     0.99 |     1.03 |     2.03 |
|------------------------------+----------+----------+----------|
|ALL                           |    43.13 |    56.87 |   100.00 |
--------------------------------------------------------------------
```

ALL

```
--------------------------------------------------------------------
|                              |            Gender         |        |
|                              |---------------------------|        |
|                              |   Male   |  Female  |   ALL   |
|                              |----------+----------+----------|
|                              |   PCTN   |   PCTN   |   PCTN   |
|------------------------------+----------+----------+----------|
|Race                          |          |          |          |
|------------------------------|          |          |          |
|White                         |    43.82 |    53.54 |    97.36 |
|------------------------------+----------+----------+----------|
|Black                         |     1.20 |     1.44 |     2.64 |
|------------------------------+----------+----------+----------|
|ALL                           |    45.02 |    54.98 |   100.00 |
--------------------------------------------------------------------
```

Percentages for Analysis Variables

Generally, when you use percentages, you want to compare various *categories* of your data. For example, you might want to know what percentage of college graduates are male versus female. However, you can also use percentages to compare two *analysis* variables.

The following code produces a table with columns for two different analysis variables: INCOME and NEWINC.

```
PROC TABULATE DATA=TEMP;
   CLASS GENDER EDUC;
   VAR INCOME NEWINC;
   TABLE GENDER*EDUC, INCOME*MEAN NEWINC*MEAN;
RUN;
```

When you look at the resulting table in Output 8.18, you can see the two income measures side by side. But the table would be more useful if you could add a column that quantified the difference between the two measures.

You could use a DATA step to create a variable that measures the difference between the two measures and then report the mean difference, but a percentage difference would be more useful.

As it happens, there is a way to do this using PROC TABULATE, so you don't have to add another DATA step. You can use PCTSUM to compare two analysis variables. All you have to do is use an asterisk operator to cross the PCTSUM statistic with the variable you want to use as the numerator, and then put the variable you want to use as the denominator in the PCTSUM denominator definition.

For our example, we'd like to show the new income measure as a percentage of the old income measure. In other words, we want to compute NEWINC/INCOME. To do this, we use an asterisk to apply a PCTSUM to NEWINC (the numerator), and we use brackets to specify INCOME as the denominator.

```
PROC TABULATE DATA=TEMP;
   CLASS GENDER EDUC;
   VAR INCOME NEWINC;
   TABLE GENDER*EDUC, INCOME*MEAN NEWINC*MEAN NEWINC*PCTSUM<INCOME>;
RUN;
```

The resulting table is shown in Output 8.19. If you check the math, you will see that the PCTSUM is equal to the MEAN of NEWINC over the MEAN of INCOME.

Output 8.18

		Old Income	New Income
		MEAN	MEAN
Gender	Education		
Male	<HS	15804.44	14953.07
	HS graduate	26627.07	26957.83
	College	43193.13	43320.97
Female	<HS	8921.97	8706.40
	HS graduate	13689.94	14091.32
	College	23493.11	23836.36

Output 8.19

		Old Income	New Income	New Income
		MEAN	MEAN	PCTSUM
Gender	Education			
Male	<HS	15804.44	14953.07	94.61
	HS graduate	26627.07	26957.83	101.24
	College	43193.13	43320.97	100.30
Female	<HS	8921.97	8706.40	97.58
	HS graduate	13689.94	14091.32	102.93
	College	23493.11	23836.36	101.46

Handling Missing Data

If data sets were perfect, you wouldn't need to read this chapter. However, in the real world, data sets have problems. One of these problems is missing data.

The TABULATE procedure is pretty good at dealing with missing data on its own. It has decision rules for what to do with missing data for classification variables and analysis variables. However, it makes these decisions without letting you know what has happened, so you may be surprised by your results.

The Example

This chapter uses a typical table to show how to find out if you have a missing data problem and to show several strategies for dealing with missing data after you've found it.

		No. Employees
		SUM
Occupation	Union Status	999
Managerial	Non-Union	999
	Union	no data
Professional	Non-Union	999
	Union	999
Technical	Non-Union	999
	Union	999
Sales	Non-Union	999
	Union	999
ALL		999

CHAPTER 9

Missing Data: How to Find It

Just because your PROC TABULATE code runs without errors and your table looks fine, don't think that there aren't problems lurking beneath the surface. Checking out your table for missing data problems is an important step in the table production process.

When you have missing data for some or all of the class variables used in your table, PROC TABULATE doesn't give you any warning messages about the problem. While many SAS procedures will produce messages like "Warning: 312 observations have missing data, only 57 observations will be used for analysis," PROC TABULATE does not.

The only way to be sure you have no problem is to look for missing data yourself. How do you do this? PROC TABULATE provides an option called MISSING. This option tells PROC TABULATE to treat missing data as a valid category for all class variables. The syntax is as follows:

```
PROC TABULATE DATA=_____  MISSING;
```

When producing a table, PROC TABULATE first checks every observation in the data set to see if it has valid data for all of the class variables. Any observations that are missing data on even one of the CLASS variables are dropped before the table is produced. To see what happens, look at the following table (run without the MISSING option):

```
PROC TABULATE DATA=TEMP;
   CLASS OCCUP UNION;
   VAR NUMEMP;
   TABLE OCCUP*UNION ALL, NUMEMP;
RUN;
```

In this table, shown in Output 9.1, you can see one form of missing data that is not hidden. In the cell marked ❶, you can see that there is no data on a number of employees for managerial union employees. This is missing data for an *analysis* variable. This type of missing data does not cause observations to be dropped from the table, only missing data on *class* variables causes observations to be dropped. To see the observations that were dropped, look at the following example, which has the MISSING option turned on:

```
PROC TABULATE DATA=TEMP MISSING;
   CLASS OCCUP UNION;
   VAR NUMEMP;
   TABLE OCCUP*UNION ALL, NUMEMP;
RUN;
```

The second table, shown in Output 9.2, looks quite different from the first. If you look at the row marked ❷, you can see that there are a number of observations where the union status for technical employees is unknown. The sales category also has some missing data. This caused a number of observations to be dropped from the table, which affects the total at the bottom of the table. Using the MISSING option makes it easy to spot this data problem. The following sections will describe various options for fixing the problem.

Output 9.1

```
----------------------------------------------
|                           |    No.    |
|                           | Employees |
|                           |-----------|
|                           |    SUM    |
|---------------------------+-----------|
|Occupation   |Union Status |           |
|-------------+-------------|           |
|Managerial   |Non-Union    |  71740.00 |
|             |-------------+-----------|
|             |Union        |       ❶ . |
|-------------+-------------+-----------|
|Professional |Non-Union    |  92458.00 |
|             |-------------+-----------|
|             |Union        |  32760.00 |
|-------------+-------------+-----------|
|Technical    |Non-Union    |   8380.00 |
|             |-------------+-----------|
|             |Union        |   5050.00 |
|-------------+-------------+-----------|
|Sales        |Non-Union    |  34506.00 |
|             |-------------+-----------|
|             |Union        |   3000.00 |
|-------------+-------------+-----------|
|ALL          |             | 247894.00 |
----------------------------------------------
```

Output 9.2

```
----------------------------------------------
|                           |    No.    |
|                           | Employees |
|                           |-----------|
|                           |    SUM    |
|---------------------------+-----------|
|Occupation   |Union Status |           |
|-------------+-------------|           |
|Managerial   |Non-Union    |  71740.00 |
|             |-------------+-----------|
|             |Union        |         . |
|-------------+-------------+-----------|
|Professional |Non-Union    |  92458.00 |
|             |-------------+-----------|
|             |Union        |  32760.00 |
|-------------+-------------+-----------|
|Technical    |.          ❷ |   2052.00 |
|             |-------------+-----------|
|             |Non-Union    |   8380.00 |
|             |-------------+-----------|
|             |Union        |   5050.00 |
|-------------+-------------+-----------|
|Sales        |.            |   6416.00 |
|             |-------------+-----------|
|             |Non-Union    |  34506.00 |
|             |-------------+-----------|
|             |Union        |   3000.00 |
|-------------+-------------+-----------|
|ALL          |             | 256362.00 |
----------------------------------------------
```

Reporting the Missing Data

The previous example showed how to spot missing data for *classification* variables in your table. But the MISSING option doesn't work for *analysis* variables. To show how many observations you have with missing data for the analysis variable, you use the NMISS statistic.

You could always use PROC MEANS to see how much missing data there is for an analysis variable, but PROC MEANS won't tell you where these values are. If you're concerned about which particular categories in your table have missing data, you need to use PROC TABULATE. With PROC TABULATE and NMISS, you can produce a table that shows which table cells have missing data for your analysis variable.

We start with a basic table that shows the number of observations for each combination of row and column classifications.

```
PROC TABULATE DATA=TEMP;
   CLASS OCCUP UNION;
   VAR NUMEMP;
   TABLE OCCUP*NUMEMP, UNION*N;
RUN;
```

In Output 9.3, you can see that one category has no nonmissing values (the cell for managerial union employees), but you can't see the other missing data. To do this, we need to switch from the N statistic to the NMISS statistic.

```
PROC TABULATE DATA=TEMP;
   CLASS OCCUP UNION;
   VAR NUMEMP;
   TABLE OCCUP*NUMEMP, UNION*NMISS;
RUN;
```

Now you can see in Output 9.4 that there is more than one category with missing data. There is scattered missing data throughout the table. This type of table makes a useful missing data report.

But you can make it even better by combining the two tables shown so far into a single table showing both the number of observations and the number of missing observations.

```
PROC TABULATE DATA=TEMP;
   CLASS OCCUP UNION;
   VAR NUMEMP;
   TABLE OCCUP*NUMEMP, UNION*(N NMISS);
RUN;
```

As you can see in Output 9.5, this table makes it easy to see how much missing data there is, and also see whether this is a problem relative to the total number of observations in the category. A cell that is missing one or two observations out of 100 is not a big problem. But a cell that is missing 50 observations out of 100 is another matter altogether.

Output 9.3

```
--------------------------------------------------------------
|                               |      Union Status          |
|                               |----------------------------|
|                               | Non-Union   |   Union      |
|                               |------------+---------------|
|                               |     N       |     N        |
|-------------------------------+------------+---------------|
|Occupation     |               |            |              |
|---------------+---------------|            |              |
|Managerial     |No. Employees  |    132.00  |      0.00    |
|---------------+---------------+------------+---------------|
|Professional   |No. Employees  |    164.00  |     52.00    |
|---------------+---------------+------------+---------------|
|Technical      |No. Employees  |     18.00  |      6.00    |
|---------------+---------------+------------+---------------|
|Sales          |No. Employees  |     69.00  |      3.00    |
--------------------------------------------------------------
```

Output 9.4

```
--------------------------------------------------------------
|                               |      Union Status          |
|                               |----------------------------|
|                               | Non-Union   |   Union      |
|                               |------------+---------------|
|                               |   NMISS     |   NMISS      |
|-------------------------------+------------+---------------|
|Occupation     |               |            |              |
|---------------+---------------|            |              |
|Managerial     |No. Employees  |      2.00  |     10.00    |
|---------------+---------------+------------+---------------|
|Professional   |No. Employees  |      2.00  |      2.00    |
|---------------+---------------+------------+---------------|
|Technical      |No. Employees  |      0.00  |      0.00    |
|---------------+---------------+------------+---------------|
|Sales          |No. Employees  |      1.00  |      0.00    |
--------------------------------------------------------------
```

Output 9.5

```
--------------------------------------------------------------------------------
|                               |              Union Status                    |
|                               |----------------------------------------------|
|                               |    Non-Union          |       Union          |
|                               |-----------------------+----------------------|
|                               |    N    |   NMISS     |    N    |   NMISS     |
|-------------------------------+---------+-------------+---------+-------------|
|Occupation     |               |         |             |         |             |
|---------------+---------------|         |             |         |             |
|Managerial     |No. Employees  | 132.00  |     2.00    |   0.00  |    10.00    |
|---------------+---------------+---------+-------------+---------+-------------|
|Professional   |No. Employees  | 164.00  |     2.00    |  52.00  |     2.00    |
|---------------+---------------+---------+-------------+---------+-------------|
|Technical      |No. Employees  |  18.00  |     0.00    |   6.00  |     0.00    |
|---------------+---------------+---------+-------------+---------+-------------|
|Sales          |No. Employees  |  69.00  |     1.00    |   3.00  |     0.00    |
--------------------------------------------------------------------------------
```

Recoding the Data

When you know that you have missing data, there are many ways you can deal with it. You can ignore it. If there are only a few missing observations, it may not be worth worrying about. But if you've got a lot of missing data, you're going to have to find a way to deal with the problem. One of the easiest ways is to recode the data.

For example, in our sample data, we have a number of observations where the union status is missing. We can recode them to 'Unknown', and then they can be included in the table with an easily understood label. This process requires a DATA step and a PROC FORMAT.

```
DATA FIXED;
   SET TEMP;
   IF UNION=. THEN UNION=0;
RUN;
PROC FORMAT;
   VALUE MISSFT  0='Unknown'
                 1='Non-Union'
                 2='Union';
RUN;
PROC TABULATE DATA=FIXED;
   CLASS OCCUP UNION;
   VAR NUMEMP;
   FORMAT UNION MISSFT.;
   TABLE OCCUP*UNION, NUMEMP;
RUN;
```

The resulting table is shown in Output 9.6. Now the union status for a subgroup of technical employees is clearly labeled as 'Unknown'. Compare this to the table in Output 9.2, where this cell is marked only with a dot. A dot may be clear to us SAS users, but it's not clear to most users. Recoding the variable creates a much more understandable table.

There is another way to recode the data. The following code uses a format for missing values and the MISSING option in the PROC TABULATE statement.

```
PROC FORMAT;
   VALUE MISSFT  .='Unknown'
                 1='Non-Union'
                 2='Union';
RUN;
PROC TABULATE DATA=FIXED MISSING;
   CLASS OCCUP UNION;
   VAR NUMEMP;
   FORMAT UNION MISSFT.;
   TABLE OCCUP*UNION, NUMEMP;
RUN;
```

The catch with this approach is that MISSING displays the results of all missing values, as you can see in Output 9.7. If you do not want to display missing values for all variables, the first approach is better.

Output 9.6

```
-----------------------------------------------
|                          |      No.    |
|                          |  Employees  |
|                          |-----------  |
|                          |     SUM     |
|--------------------------+-----------  |
|Occupation   |Union Status |             |
|-------------+-------------|             |
|Managerial   |Non-Union    |   71740.00  |
|             |-------------+-----------  |
|             |Union        |          .  |
|-------------+-------------+-----------  |
|Professional |Non-Union    |   92458.00  |
|             |-------------+-----------  |
|             |Union        |   32760.00  |
|-------------+-------------+-----------  |
|Technical    |Unknown      |    2052.00  |
|             |-------------+-----------  |
|             |Non-Union    |    8380.00  |
|             |-------------+-----------  |
|             |Union        |    5050.00  |
|-------------+-------------+-----------  |
|Sales        |Unknown      |    6416.00  |
|             |-------------+-----------  |
|             |Non-Union    |   34506.00  |
|             |-------------+-----------  |
|             |Union        |    3000.00  |
-----------------------------------------------
```

Output 9.7

```
-----------------------------------------------
|                          |      No.    |
|                          |  Employees  |
|                          |-----------  |
|                          |     SUM     |
|--------------------------+-----------  |
|Occupation   |Union        |             |
|-------------+-------------|             |
|.            |Union        |    2050.00  |
|-------------+-------------+-----------  |
|Managerial   |Non-Union    |   71740.00  |
|             |-------------+-----------  |
|             |Union        |          .  |
|-------------+-------------+-----------  |
|Professional |Non-Union    |   92458.00  |
|             |-------------+-----------  |
|             |Union        |   32760.00  |
|-------------+-------------+-----------  |
|Technical    |Unknown      |    2052.00  |
|             |-------------+-----------  |
|             |Non-Union    |    8380.00  |
|             |-------------+-----------  |
|             |Union        |    5050.00  |
|-------------+-------------+-----------  |
|Sales        |Unknown      |    6416.00  |
|             |-------------+-----------  |
|             |Non-Union    |   34506.00  |
|             |-------------+-----------  |
|             |Union        |    3000.00  |
-----------------------------------------------
```

Formatting Missing Values

In the last example, we recoded the values of a *classification* variable to label the rows with missing data as 'Unknown.' You can also recode the values displayed in table cells for *analysis* variables with missing data.

You find the "." missing data code in table cells when there is missing data for the analysis variable in every observation. PROC TABULATE gives you the option to replace the "." with some other text of your choice. This makes the table more understandable.

For example, take a look at the following table:

```
PROC TABULATE DATA=TEMP;
   CLASS OCCUP UNION;
   VAR NUMEMP;
   TABLE OCCUP*UNION ALL, NUMEMP;
RUN;
```

In Output 9.8, you can see that the table has no data for the number of union managerial employees. This missing data is marked with a ".". We can change this by using the PROC TABULATE option MISSTEXT. This option is used on the TABLE statement to specify what text PROC TABULATE should use in place of the "." marker.

The MISSTEXT option is added by using a slash after the row and column definitions. In the following code, MISSTEXT is set to the text 'no data'.

```
PROC TABULATE DATA=TEMP;
   CLASS OCCUP UNION;
   VAR NUMEMP;
   TABLE OCCUP*UNION ALL, NUMEMP / MISSTEXT='no data';
RUN;
```

The resulting table is shown in Output 9.9. Notice how the cell that used to have a "." now has the words 'no data'. This makes the table much clearer to nontechnical end users.

This technique will not work for classification variables. If you run the same code with the MISSING option turned on, any classification variables with missing data will still show up in the row and column headers marked with a ".".

So if you have a table with missing data both for classification variables and for analysis variables, and you want to recode both types of missing data from "." to 'Unknown' or 'no data,' you need to take two steps. First, recode the missing data for classification variables to a nonmissing value, and then format the results as desired. Second, use the MISSTEXT option to format the missing data for analysis variables.

Output 9.8

```
----------------------------------------------------
|                             |      No.    |
|                             |  Employees  |
|                             | ----------- |
|                             |     SUM     |
| ----------------------------+------------ |
|Occupation    |Union Status  |             |
| -------------+------------- |             |
|Managerial    |Non-Union     |   71740.00  |
|              | -------------+------------ |
|              |Union         |           . |
| -------------+-------------+------------- |
|Professional  |Non-Union     |   92458.00  |
|              | -------------+------------ |
|              |Union         |   32760.00  |
| -------------+-------------+------------- |
|Technical     |Non-Union     |    8380.00  |
|              | -------------+------------ |
|              |Union         |    5050.00  |
| -------------+-------------+------------- |
|Sales         |Non-Union     |   34506.00  |
|              | -------------+------------ |
|              |Union         |    3000.00  |
| ---------------------------+------------- |
|ALL                          |  247894.00  |
----------------------------------------------------
```

Output 9.9

```
----------------------------------------------------
|                             |      No.    |
|                             |  Employees  |
|                             | ----------- |
|                             |     SUM     |
| ----------------------------+------------ |
|Occupation    |Union Status  |             |
| -------------+------------- |             |
|Managerial    |Non-Union     |   71740.00  |
|              | -------------+------------ |
|              |Union         |    no data  |
| -------------+-------------+------------- |
|Professional  |Non-Union     |   92458.00  |
|              | -------------+------------ |
|              |Union         |   32760.00  |
| -------------+-------------+------------- |
|Technical     |Non-Union     |    8380.00  |
|              | -------------+------------ |
|              |Union         |    5050.00  |
| -------------+-------------+------------- |
|Sales         |Non-Union     |   34506.00  |
|              | -------------+------------ |
|              |Union         |    3000.00  |
| -------------+-------------+------------- |
|ALL                          |  247894.00  |
----------------------------------------------------
```

Three-Dimensional Tables with Missing Data

If you are building a three-dimensional table, then you have an additional option for handling missing data. Because it can be useful to have every page of a three-dimensional table look alike, PROC TABULATE gives you the option to force them to look alike.

You can force a table to show a classification variable that has no nonmissing values for an entire row or column. In a two-dimensional table, any classifications that have no nonmissing data are dropped from the table. With the PRINTMISS option, you can force a three-dimensional table to show these values. As long as the classification has nonmissing data on *one* of the pages, it will be shown on *all* of the pages.

For example, the following code produces a two-page table:

```
PROC TABULATE DATA=TEMP;
   CLASS UNION OCCUP GENDER;
   VAR NUMEMP;
   TABLE GENDER, UNION, OCCUP*NUMEMP*SUM;
RUN;
```

If you look at the output in Output 9.10, you can see that the second page of the table does not match the first page of the table. Because there is no data on union status for female sales employees, that column is dropped from the table. This can be confusing to readers of the table, who will wonder why you dropped one column from one page to the next.

You can fix this problem by specifying the PRINTMISS option. This is a TABLE statement option, so you add it with a slash following the TABLE definitions.

```
PROC TABULATE DATA=TEMP;
   CLASS UNION OCCUP GENDER;
   VAR NUMEMP;
   TABLE GENDER, UNION, OCCUP*NUMEMP*SUM
      / PRINTMISS;
RUN;
```

The new table is shown in Output 9.11. The first page of the table has been omitted because it is identical to the table for males shown in Output 9.10 above. The second page of the table is what we're interested in. Notice how it now matches the first page in width. All four occupations are represented, even though there is no data for the Sales column.

What PROC TABULATE has done is to treat missing values as a legal CLASS variable value, even though the entire classification is missing on this page. This is different from the MISSING option because the MISSING option will not show the missing data if there are no nonmissing values for the entire classification.

In general, you should specify PRINTMISS when you build three-dimensional tables. It makes your tables uniform and easy to read. The only time you might want to drop this option is if you are building huge tables with lots of missing data. In these cases, the space savings from omitting some rows and columns may be worth the loss in table consistency.

Output 9.10

```
Gender Male
-----------------------------------------------------------------------------
|                 |                         Occupation                      |
|                 |-------------------------------------------------------- |
|                 | Managerial  | Professional | Technical   |  Sales       |
|                 |-------------+--------------+-------------+-------------- |
|                 |No. Employees|No. Employees |No. Employees|No. Employees |
|                 |-------------+--------------+-------------+-------------- |
|                 |    SUM      |    SUM       |    SUM       |    SUM       |
|-----------------+-------------+--------------+-------------+------------- |
|Union Status     |             |              |             |              |
|---------------- |             |              |             |              |
|Non-Union        |    42986.00 |    56286.00  |    4900.00   |   11190.00   |
|-----------------+-------------+--------------+-------------+------------- |
|Union            |          .  |    10810.00  |    2000.00   |    1000.00   |
-----------------------------------------------------------------------------
```

```
Gender Female
------------------------------------------------------------------
|                 |                   Occupation                  |
|                 |---------------------------------------------- |
|                 | Managerial  | Professional | Technical        |
|                 |-------------+--------------+---------------    |
|                 |No. Employees|No. Employees |No. Employees     |
|                 |-------------+--------------+---------------    |
|                 |    SUM      |    SUM       |    SUM            |
|-----------------+-------------+--------------+---------------    |
|Union Status     |             |              |                   |
|---------------- |             |              |                   |
|Non-Union        |    28754.00 |    36172.00  |    3480.00        |
|-----------------+-------------+--------------+---------------    |
|Union            |          .  |    21950.00  |    3050.00        |
------------------------------------------------------------------
```

Output 9.11

(Table for Gender=Male is not shown. It is the same as shown in Output 9.10 above.)

```
Gender Female
-----------------------------------------------------------------------------
|                 |                         Occupation                      |
|                 |-------------------------------------------------------- |
|                 | Managerial  | Professional | Technical   |  Sales       |
|                 |-------------+--------------+-------------+-------------- |
|                 |No. Employees|No. Employees |No. Employees|No. Employees |
|                 |-------------+--------------+-------------+-------------- |
|                 |    SUM      |    SUM       |    SUM       |    SUM       |
|-----------------+-------------+--------------+-------------+------------- |
|Union Status     |             |              |             |              |
|---------------- |             |              |             |              |
|Non-Union        |    28754.00 |    36172.00  |    3480.00   |         .    |
|-----------------+-------------+--------------+-------------+------------- |
|Union            |          .  |    21950.00  |    3050.00   |         .    |
-----------------------------------------------------------------------------
```

Breaking Up the Table

PROC TABULATE drops all observations that have missing data on *any* of the CLASS variables in your table. This means that if you have one classification with a lot of missing data, you'll lose these observations for all of the other classification variables as well. Even if the other classifications have valid data, these observations will be dropped from the entire table.

For example, in the following table, the MISSING option is used to spot a large cluster of missing data for the variable NUMKIDS:

```
PROC TABULATE DATA=TEMP MISSING;
   CLASS MARITAL NUMKIDS EDUC;
   TABLE EDUC ALL, (MARITAL NUMKIDS)*N;
RUN;
```

In Output 9.12, you can see that there are 251 observations with missing data for a number of children. When we remove the MISSING option to produce a clean final report, we will lose these 251 observations not just from the columns for number of children but also from the columns for marital status. But we know that these observations have valid data for marital status because there's no missing data column showing on that side of the table.

There are two options at this point. First, we could leave the MISSING option on. This would leave an ugly missing data column in the final report. Second, we could break up the table. If we run each half of the columns as a separate table, then we can use all of the available observations for the marital status table.

```
PROC TABULATE DATA=TEMP;
   CLASS MARITAL EDUC;
   TABLE EDUC ALL, MARITAL*N;
RUN;
PROC TABULATE DATA=TEMP;
   CLASS NUMKIDS EDUC;
   TABLE EDUC ALL, NUMKIDS*N;
RUN;
```

As you can see in Output 9.13, now both tables maximize their use of the data, and there is no column of missing data in the number of children table. In general, if you have missing data, it's better to break up your tables so that missing data for some classification variables does not cause a large number of observations to be dropped.

By the way, it is possible for missing data for variables not even used in your table to cause observations to be dropped. Watch out for extra variables in your CLASS statement that aren't used in the TABLE statement. PROC TABULATE drops all observations with missing data for any of the CLASS variables, whether or not they are actually used in the table. It's easy to forget when you drop a classification from the table to also drop the variable from the CLASS statement. (Extra variables in the VAR statement do not have this affect.)

Output 9.12

		Marital Status				No. of Children		
		Married	Widow/ Div/Sep	Never Married	.	0	1	2+
		N	N	N	N	N	N	N
Education								
<HS		526.00	319.00	156.00	33.00	210.00	247.00	511.00
HS graduate		1486.00	508.00	288.00	65.00	514.00	527.00	1176.00
College		2245.00	507.00	604.00	153.00	690.00	898.00	1615.00
ALL		4257.00	1334.00	1048.00	251.00	1414.00	1672.00	3302.00

Output 9.13

	Marital Status		
	Married	Widow/ Div/Sep	Never Married
	N	N	N
Education			
<HS	526.00	319.00	156.00
HS	1486.00	508.00	288.00
College	2245.00	507.00	604.00
ALL	4257.00	1334.00	1048.00

	No. of Children		
	0	1	2+
	N	N	N
Education			
<HS	210.00	247.00	511.00
HS graduate	514.00	527.00	1176.00
College	690.00	898.00	1615.00
ALL	1414.00	1672.00	3302.00

DATA Step Tricks

What happens when you build a table to show data for a variable that has three different possible values, say 'A', 'B', and 'C', but in your data set, you end up with records for 'A' and 'B' only? This situation often comes up when you are dealing with survey data. A question had three valid responses, but the respondents only chose two of the three answers. It would be nice if the PROC TABULATE table would show all three choices, even though one was never chosen.

You can't do this with the MISSING option. That option shows only that you have records in the data set where there is missing data. In this case we're not missing individual values, we're missing records. There are plenty of records where 'A' and 'B' were chosen, but there are no records where 'C' was chosen.

In the example table shown in Output 9.14, you can see that there are two sectors represented: 'Private' and 'Public'. What you can't tell from the table is that on the survey from which this data was gathered, there was a third choice for sector: 'Non-Profit'. It just happened that none of the respondents selected that item.

To keep the table consistent with tables from previous surveys, it would be nice to be able to force a column for 'Non-Profit' to appear. The way to do this is to use the DATA step to create some fake data. What we need is one record for each of the four occupation categories that has SECTOR set to 'Non-Profit'. To keep these records from adding false counts to the number of employees shown in the table, we set that variable to zero for each of these fake records. The DATA step and PROC TABULATE code are shown below.

```
DATA FIXIT;
   SET TEMP;
   OUTPUT;
   IF _N_=1 THEN DO N=1 TO 4;
      OCCUP=N;
      SECTOR='Non-Profit';
      NUMEMP=0;
      OUTPUT;
   END;
RUN;
PROC TABULATE DATA=FIXIT;
   CLASS OCCUP SECTOR;
   VAR NUMEMP;
   TABLE OCCUP ALL, SECTOR*NUMEMP;
RUN;
```

You can see from the table shown in Output 9.15 that we now have a column showing for 'Non-Profit'. It shows a total of zero employees for each occupation group.

This is a solution that will not work in all situations. In this case, we had an analysis variable that would be meaningful if set to zero. If we had a different sort of analysis variable or if we had chosen a different statistic, the table would be meaningless. For example, if the table were displaying mean incomes, we wouldn't want to show the nonprofit sector with a mean income of $0!

Output 9.14

		SECTOR	
		Private	Public
		No. Employees	No. Employees
		SUM	SUM
Occupation			
Managerial		58046.00	20796.00
Professional		88398.00	36820.00
Technical		12482.00	3000.00
Sales		42922.00	1000.00
ALL		201848.00	61616.00

Output 9.15

		SECTOR		
		Non-Profit	Private	Public
		No. Employees	No. Employees	No. Employees
		SUM	SUM	SUM
Occupation				
Managerial		0.00	58046.00	20796.00
Professional		0.00	88398.00	36820.00
Technical		0.00	12482.00	3000.00
Sales		0.00	42922.00	1000.00
ALL		0.00	201848.00	61616.00

Version 7 provides better tools to solve this type of problem. You can specify that all levels of a format should be displayed, whether or not they have data (the PRELOADFMT and EXCLUSIVE options), or you can specify a data set that contains the values you want included, whether or not they have data (CLASSDATA and EXCLUSIVE options). See Chapter 24, "Version 7 Enhancements," for more information.

Modifying Row and Column Headings

By now you're probably thinking that the tables shown in this book are pretty ugly. At this point, we've got all of the basics covered and we can turn to the finer points of table construction. The next four chapters are devoted to cleaning up, reorganizing, and otherwise beautifying your PROC TABULATE tables.

The Example

This chapter explains a number of techniques you can use to make your row and column headings more concise and attractive. We'll relabel variables and statistics to make them clearer. Instead of row and column headings that are stacked three and four items deep, we'll get rid of the excess and produce more readable tables, as in the following example:

Income		Employed Full-Time		Employed Part-Time		Total	
		Mean	Std Dev	Mean	Std Dev	Mean	Std Dev
<HS		999	999	999	999	999	999
	Males	999	999	999	999	999	999
	Females	999	999	999	999	999	999
	Total	999	999	999	999	999	999
...		999	999	999	999	999	999

(This table has been truncated to fit on the page. There are additional rows for education levels HS and College.)

Modifying Variable Labels

When we put variables in our tables, PROC TABULATE adds labels for the appropriate rows or columns that indicate which variable is being displayed where. But variable names are not always very descriptive. The variable name OCCUP may be clear to you, but people reading your table may be confused by this cryptic label. A better approach is to use a descriptive label for your variable. There are two ways to do this.

The first way to add labels to your variables is by using a LABEL statement just like you can in any other procedure. In the following example, three of the variables in the table are assigned new labels. The first two, OCCUP and FULLTIME, are assigned their labels using an ordinary LABEL statement. The third variable, ALL, is a special variable and needs special handling. Because ALL does not exist in the data set TEMP, it cannot be labeled in the usual fashion. Instead, we need to use the KEYLABEL statement. This statement allows you to add labels to TABULATE keywords.

```
PROC TABULATE DATA=TEMP;
   CLASS OCCUP FULLTIME;
   VAR INCOME;
   TABLE OCCUP ALL, (FULLTIME ALL)*INCOME*MEAN;
   LABEL OCCUP='Occupation'
         FULLTIME='Employment Status';
   KEYLABEL ALL='TOTAL';
RUN;
```

The resulting table is shown in Output 10.1. Notice how OCCUP and FULLTIME now have more understandable labels. And also notice how 'ALL' has been replaced by 'TOTAL' in both the row and column headings. This is the advantage of the KEYLABEL statement. It replaces every occurrence of a keyword with its label.

But this can be a disadvantage at times. What if you want to use one label for ALL for the rows and one label for ALL for the columns? You can't use a LABEL or KEYLABEL statement unless the variable or keyword can have the same label everywhere. To get around this, you need to change the variable labels individually in the TABLE statement.

In the following example, each of the labels is assigned in the TABLE statement by using an equal sign after the variable name and then putting the label in quotation marks. The equal sign is another PROC TABULATE *operator,* which is used to apply a label to a variable. The labels for OCCUP and FULLTIME are modified just as they were in the previous example. But if you look at the two occurrences of ALL, you can see that the row totals will be labeled 'All Occupations' and the column totals will be labeled 'Total'.

```
PROC TABULATE DATA=TEMP;
   CLASS OCCUP FULLTIME;
   VAR INCOME;
   TABLE OCCUP='Occupation' ALL='All Occupations',
      (FULLTIME='Employment Status' ALL='Total')*INCOME*MEAN;
RUN;
```

The resulting table is shown in Output 10.2. As you can see, the two tables produced in this example aren't that different. Unless you need the flexibility of TABLE statement labeling, it really doesn't matter which way you assign your labels. Use the method that is easiest for you to understand.

Output 10.1

```
---------------------------------------------------------------------
|                              | Employment Status |               |
|                              |-------------------|               |
|                              | Full time | Part time |  TOTAL    |
|                              |-----------+-----------+---------- |
|                              | Income    | Income    | Income    |
|                              |-----------+-----------+---------- |
|                              | MEAN      | MEAN      | MEAN      |
|------------------------------+-----------+-----------+---------- |
|Occupation                    |           |           |           |
|------------------------------|           |           |           |
|Managerial                    | 47300.70| 37303.41|  46306.34 |
|------------------------------+-----------+-----------+---------- |
|Professional                  | 40807.71| 41894.13|  40923.65 |
|------------------------------+-----------+-----------+---------- |
|Technical                     | 31028.56| 31722.50|  31111.42 |
|------------------------------+-----------+-----------+---------- |
|Sales                         | 33023.21| 27997.97|  32547.52 |
|------------------------------+-----------+-----------+---------- |
|TOTAL                         | 40961.30| 37103.26|  40565.32 |
---------------------------------------------------------------------
```

Output 10.2

```
---------------------------------------------------------------------
|                              | Employment Status |               |
|                              |-------------------|               |
|                              | Full time | Part time |  Total    |
|                              |-----------+-----------+---------- |
|                              | Income    | Income    | Income    |
|                              |-----------+-----------+---------- |
|                              | MEAN      | MEAN      | MEAN      |
|------------------------------+-----------+-----------+---------- |
|Occupation                    |           |           |           |
|------------------------------|           |           |           |
|Managerial                    | 47300.70| 37303.41|  46306.34 |
|------------------------------+-----------+-----------+---------- |
|Professional                  | 40807.71| 41894.13|  40923.65 |
|------------------------------+-----------+-----------+---------- |
|Technical                     | 31028.56| 31722.50|  31111.42 |
|------------------------------+-----------+-----------+---------- |
|Sales                         | 33023.21| 27997.97|  32547.52 |
|------------------------------+-----------+-----------+---------- |
|All Occupations               | 40961.30| 37103.26|  40565.32 |
---------------------------------------------------------------------
```

Unlike PROC PRINT, PROC TABULATE does not have a SPLIT= option, so if you have variables with long labels containing split characters that indicate where to put line breaks, you'll need to remove these characters and come up with shorter variable labels to use in PROC TABULATE.

Modifying Statistic Labels

When we put statistics in our tables, PROC TABULATE adds labels for the appropriate rows or columns that indicate which statistic has been selected. As SAS programmers, the labels N, MEAN, STD, and RANGE may make perfect sense. But what about the end users?

Wouldn't it be nice if you could add more descriptive labels? As it happens, you can. PROC TABULATE supplies default labels for statistics, but they can be renamed to anything you want. For example, in the following table, we display three statistics. The following code generates a table with the default labels:

```
PROC TABULATE DATA=TEMP;
   CLASS OCCUP FULLTIME;
   VAR INCOME;
   TABLE OCCUP, FULLTIME*INCOME*(N MEAN STD);
RUN;
```

The results are shown in Output 10.3. There's nothing wrong with this table, it displays the required results. But it would be nice to make clearer labels for the statistics. For example, it would be nice to label 'N' as 'NUMBER OF OBS' and STD as 'STANDARD DEVIATION'.

The way you change the label of a PROC TABULATE statistic is to follow the statistic keyword with an equal sign and then the new label in quotes. You can label statistics in the TABLE statement just like you label variables in the TABLE statement. So instead of listing N in the TABLE statement, we list N='NUMBER OF OBS'. And instead of listing STD, we list STD='STANDARD DEVIATION'. The MEAN keyword is left alone because it was clear enough as is.

```
PROC TABULATE DATA=TEMP;
   CLASS OCCUP FULLTIME;
   VAR INCOME;
   TABLE OCCUP, FULLTIME*INCOME*
      (N='NUMBER OF OBS.' MEAN STD='STANDARD DEVIATION');
RUN;
```

The modified table is shown in Output 10.4. As you can see, the labels for N and STD have been replaced with their new labels. The label for MEAN was left unchanged. Notice that because the new labels are longer than the originals, PROC TABULATE makes the spaces for the labels two lines deep. PROC TABULATE will make room for whatever labels you create, but you should keep labels as short and simple as possible. If you use words that are longer than the column is wide, PROC TABULATE will have to wrap the label mid-word. PROC TABULATE will *not* make column headings wider to fit new labels.

In this example, the new labels were all in upper case, just like the default labels. But you are not restricted to upper case. You can make the labels even more readable by using mixed case. For example, the previous TABLE statement could be modified as follows:

```
TABLE OCCUP, FULLTIME*INCOME*
   (N='Number of Obs.' MEAN='Mean' STD='Standard Deviation');
```

Output 10.3

	Employment					
	Full time			Part time		
	Income			Income		
	N	MEAN	STD	N	MEAN	STD
Occupation						
Managerial	670.00	47300.70	33175.97	74.00	37303.41	26220.95
Professional	745.00	40807.71	25798.11	89.00	41894.13	24398.82
Technical	118.00	31028.56	17032.01	16.00	31722.50	22091.34
Sales	373.00	33023.21	27384.35	39.00	27997.97	28592.59

Output 10.4

	Employment					
	Full time			Part time		
	Income			Income		
	NUMBER OF OBS.	MEAN	STANDARD DEVIATION	NUMBER OF OBS.	MEAN	STANDARD DEVIATION
Occupation						
Managerial	670.00	47300.70	33175.97	74.00	37303.41	26220.95
Professional	745.00	40807.71	25798.11	89.00	41894.13	24398.82
Technical	118.00	31028.56	17032.01	16.00	31722.50	22091.34
Sales	373.00	33023.21	27384.35	39.00	27997.97	28592.59

Another Way to Modify Statistic Labels

As is often the case with SAS, there is more than one way to do things. The previous example showed how to modify a statistic's label by assigning it a new label in the TABLE statement. You can also assign a new label to a statistic in a separate statement.

For PROC TABULATE statistics keywords, you can't use a LABEL statement because these are not variables. Instead, you use a KEYLABEL statement. The syntax is the same as a LABEL statement.

```
KEYLABEL    MEAN='This is a mean'
            STD='This is a standard deviation';
```

We can use this statement to reproduce the table from the previous example. In the following code, the statistic labels have been removed from the TABLE statement and placed in a KEYLABEL statement instead.

```
PROC TABULATE DATA=TEMP;
    CLASS OCCUP FULLTIME;
    VAR INCOME;
    TABLE OCCUP, FULLTIME*INCOME*(N MEAN STD);
    KEYLABEL    N='NUMBER OF OBS.'
                STD='STANDARD DEVIATION';
RUN;
```

The results are shown in Output 10.5. If you compare this table to Output 10.4, you can see that it makes no difference whether the labels are placed in the TABLE statement or in a separate KEYLABEL statement.

So why bother showing two ways to do the same thing? The answer is that there are times where the KEYLABEL statement comes in handy. Sometimes you have the same statistic keyword used in a number of places in the same table. If you modified the labels in the TABLE definition, you'd have to modify every occurrence of the statistic. By using KEYLABEL, you can change them all at once.

For example, the following table uses the MEAN statistic in more than one place, but it only needs to be re-labeled once if we use a KEYLABEL statement.

```
PROC TABULATE DATA=TEMP;
    CLASS EDUC RACE FULLTIME;
    VAR AGE;
    TABLE EDUC*(N MEAN) FULLTIME*(NMISS MEAN),
        RACE*AGE;
    KEYLABEL MEAN='Average';
RUN;
```

The results are shown in Output 10.6. Notice how every occurrence of MEAN has been changed to 'Average'.

Output 10.5

	Employment					
	Full time			Part time		
	Income			Income		
	NUMBER OF OBS.	MEAN	STANDARD DEVIATION	NUMBER OF OBS.	MEAN	STANDARD DEVIATION
Occupation						
Managerial	670.00	47300.70	33175.97	74.00	37303.41	26220.95
Professional	745.00	40807.71	25798.11	89.00	41894.13	24398.82
Technical	118.00	31028.56	17032.01	16.00	31722.50	22091.34
Sales	373.00	33023.21	27384.35	39.00	27997.97	28592.59

Output 10.6

		Race	
		White	Black
		Age	Age
Education			
<HS	N	282.00	11.00
	Average	47.77	41.91
HS	N	1306.00	25.00
	Average	42.40	41.08
College	N	2388.00	59.00
	Average	41.32	38.85
Employment			
Full time	NMISS	5.00	3.00
	Average	42.03	39.02
Part time	NMISS	1.00	2.00
	Average	42.88	45.64

Hiding Statistic Labels

The next area of cleanup we'll tackle is excessive statistic labels. Every time you display a statistic in a table, by default PROC TABULATE generates a row or column heading for that statistic.

If you've got two nested row variables and then add a statistic to the row dimension of the table, you get row headings that are three items deep. Similarly, if you have two nested column headings, an analysis variable, and then add a statistic, you'll end up with column headings that are six items deep!

In the example below, we have a table with a single row variable and two nested column variables. The statistic we have opted to display is the mean of the analysis variable income.

```
PROC TABULATE DATA=TEMP;
   CLASS YEAR FULLTIME GENDER;
   VAR INCOME;
   TABLE YEAR, FULLTIME*GENDER*INCOME*MEAN;
RUN;
```

As you can see from Output 10.7, the column headings on this table are stacked so deep it's almost impossible to figure out what's going on. As a first step to simplify the table, we can get rid of one layer of the headings.

One way to do this is to move the MEAN statistic to the row dimension, as in the following TABLE statement:

```
TABLE YEAR*MEAN, FULLTIME*GENDER*INCOME;
```

But this approach would make the rows harder to read. Instead, why don't we move the statistic label to the BOX. There's plenty of room there.

Adding the label "MEAN" to the BOX is easy. We've done this in previous examples. Just add BOX='MEAN' to the TABLE statement.

```
TABLE YEAR, FULLTIME*GENDER*INCOME*MEAN / BOX='MEAN';
```

If we ran this as is, we'd have two labels for our statistic, one in the BOX and one in the column headings. So now what we have to do is remove the extra label. We do this by assigning a blank label to the mean.

```
TABLE YEAR, FULLTIME*GENDER*INCOME*MEAN=' ' / BOX='MEAN';
```

This tells PROC TABULATE that we do not want any label for the MEAN statistic. Instead of displaying an empty box in the column heading where the "MEAN" would have gone, PROC TABULATE is smart enough to figure out that the space is not needed.

In Output 10.8, you can see that now the table has only five column headings instead of the original four. This makes the table easier to read. If we wanted to clean up the table even more, we could change the BOX label from 'MEAN' to something like 'Figures represent means'.

Output 10.7

Year	Employment			
	Full time		Part time	
	Gender		Gender	
	Male	Female	Male	Female
	Income	Income	Income	Income
	MEAN	MEAN	MEAN	MEAN
1992	49392.40	28039.92	51630.17	29574.75
1993	52752.70	33508.60	45429.57	41739.79
1994	63900.40	39023.33	49267.11	31991.33
1995	18271.82	14412.87	21853.00	8768.00
1996	23874.32	14238.68	13465.33	16124.00
1997	25888.26	16886.75	30150.00	14804.00
1998	30819.46	18181.18	28239.16	17683.69

Output 10.8

MEAN	Employment			
	Full time		Part time	
	Gender		Gender	
	Male	Female	Male	Female
	Income	Income	Income	Income
Year				
1992	49392.40	28039.92	51630.17	29574.75
1993	52752.70	33508.60	45429.57	41739.79
1994	63900.40	39023.33	49267.11	31991.33
1995	18271.82	14412.87	21853.00	8768.00
1996	23874.32	14238.68	13465.33	16124.00
1997	25888.26	16886.75	30150.00	14804.00
1998	30819.46	18181.18	28239.16	17683.69

Hiding Variable Labels

We're going to continue to work with the table from the previous example. Even though we reduced the number of column headings, there are still too many headings. There's another layer of headings that we can remove.

Because the values of GENDER are 'Male' and 'Female', it is superfluous to label the variable as 'Gender'. We can get rid of this label in the headings. You get rid of variable labels the same way that we got rid of the statistic labels in the previous example–you assign them to a blank label in the TABLE statement.

```
PROC TABULATE DATA=TEMP;
   CLASS YEAR FULLTIME GENDER;
   VAR INCOME;
   TABLE YEAR, FULLTIME*GENDER=' '*INCOME*MEAN=' '
      / BOX='Mean';
RUN;
```

In the resulting table shown in Output 10.9, you can see that we now have one less layer of headings over each column. And the table is still perfectly understandable. But we can do even better. Each column still has four headings, so why don't we get rid of some more.

We can get rid of the INCOME headings the same way we got rid of the headings for MEAN. We can change the BOX label to 'Mean Income' and then remove the column heading by assigning it a blank label. We can also get rid of the row heading, 'Year'. The values of the row variable make it perfectly clear that they represent years.

But getting rid of the heading used by the label for FULLTIME is a little harder. If we just set FULLTIME=' ' in the TABLE statement, the output may be hard to understand. 'Part time' and 'Full time' may not be clear enough. To fix this, we use a new format for the variable that gives each value a clearer label. Then we can dispense with the variable label.

```
PROC FORMAT;
   VALUE FULLTIME 1='Employed Full-Time'
                  2='Employed Part-Time';
RUN:
PROC TABULATE DATA=TEMP;
   CLASS YEAR FULLTIME GENDER;
   VAR INCOME;
   TABLE YEAR=' ',
      FULLTIME=' '*GENDER=' '*INCOME=' '*MEAN=' '
      / BOX='Mean Income';
   FORMAT FULLTIME FULLTIME.;
RUN;
```

Notice how the resulting table in Output 10.10 is much simpler and easier to read than the table we started out with back in Output 10.7. As you build complex PROC TABULATE tables, you will find yourself assigning blank labels a lot to keep the row and column headings from growing out of control.

Output 10.9

Mean	Employment			
	Full time		Part time	
	Male	Female	Male	Female
	Income	Income	Income	Income
Year				
1992	49392.40	28039.92	51630.17	29574.75
1993	52752.70	33508.60	45429.57	41739.79
1994	63900.40	39023.33	49267.11	31991.33
1995	18271.82	14412.87	21853.00	8768.00
1996	23874.32	14238.68	13465.33	16124.00
1997	25888.26	16886.75	30150.00	14804.00
1998	30819.46	18181.18	28239.16	17683.69

Output 10.10

Mean Income	Employed Full-Time		Employed Part-Time	
	Male	Female	Male	Female
1992	49392.40	28039.92	51630.17	29574.75
1993	52752.70	33508.60	45429.57	41739.79
1994	63900.40	39023.33	49267.11	31991.33
1995	18271.82	14412.87	21853.00	8768.00
1996	23874.32	14238.68	13465.33	16124.00
1997	25888.26	16886.75	30150.00	14804.00
1998	30819.46	18181.18	28239.16	17683.69

You can't assign blank labels using a LABEL statement; you have to do this in the TABLE statement. If you use a statement such as LABEL GENDER=' ' in your code, PROC TABULATE will revert to the variable name GENDER when it builds the table.

Modifying Row Headings

There is one limitation you need to keep in mind when turning variable and statistic labels into blank labels to remove excess headings. If you use blank labels for column headings, PROC TABULATE removes the excess headings and does not display a blank heading. However, for rows it's a different story.

If you have a statistic or analysis variable in the row heading and you give it a blank label, PROC TABULATE is going to allot space for it anyway. You will end up with an extra row heading that's completely empty.

For example, look at the following table. A number of the row and column variables have been relabeled as blank to save space. In particular, note that the MEAN statistic and the analysis variable INCOME in the row dimension have been given a blank label.

```
PROC TABULATE DATA=TEMP;
   CLASS EDUC GENDER FULLTIME;
   VAR INCOME;
   TABLE EDUC='Education'*GENDER=' '*INCOME=' '*MEAN=' ',
      FULLTIME=' ' / BOX='Mean Income';
RUN;
```

In the results shown in Output 10.11, you can see that the labels for MEAN and INCOME in the rows are indeed blank. However, you didn't save any space because PROC TABULATE left two empty boxes in the table grid for the labels anyway. To get rid of this extra space, you have two options.

First, you could move the statistic and the analysis variable to the column definition, where the blank label would be handled properly and the extra heading would be deleted completely. The drawback here is that sometimes you want to be able to put the statistic or the analysis variable in the row dimension.

The second option is to tell PROC TABULATE how to handle the blank row headings. You do this with the ROW= option. This option is set by default to ROW=CONSTANT, which allows equal space in the row heading for all labels of variables, class values, and statistics keywords. The other option, which we use, is ROW=FLOAT. This option allows equal space in the row heading for all *nonblank* labels for variables, class values, and statistics keywords.

```
PROC TABULATE DATA=TEMP;
   CLASS EDUC GENDER FULLTIME;
   VAR INCOME;
   TABLE EDUC='Education'*GENDER=' '*INCOME=' '*MEAN=' ',
      FULLTIME=' ' / BOX='Mean Income' ROW=FLOAT;
RUN;
```

Now if you look at Output 10.12, you will see that the blank headings are gone, and PROC TABULATE has used this extra space to provide more room for the other variables. This is the best way to handle blank row headings. In fact, you will want to keep ROW=FLOAT as a standard part of your PROC TABULATE code. If you have no blank row headings, it does no harm; and if you do have blank headings, you'll want the option turned on.

Output 10.11

```
-------------------------------------------------------------
|Mean Income                          | Employed  | Employed  |
|                                     | Full-Time | Part-Time |
|-------------------------------+-----------+-----------|
|Educat-|       |       |       |           |           |
|ion    |       |       |       |           |           |
|-------+-------+-------+-------|           |           |
|<HS    |Male   |       |       | 22625.90| 25060.91|
|       |-------+-------+-------+-----------+-----------|
|       |Female |       |       | 14850.22| 14449.73|
|-------+-------+-------+-------+-----------+-----------|
|HS     |Male   |       |       | 31085.59| 28307.08|
|       |-------+-------+-------+-----------+-----------|
|       |Female |       |       | 18189.30| 17574.26|
|-------+-------+-------+-------+-----------+-----------|
|COLLEGE|Male   |       |       | 47960.27| 42817.40|
|       |-------+-------+-------+-----------+-----------|
|       |Female |       |       | 27745.10| 28767.81|
-------------------------------------------------------------
```

Output 10.12

```
-------------------------------------------------------------
|Mean Income                          | Employed  | Employed  |
|                                     | Full-Time | Part-Time |
|-----------------------------+-----------+-----------|
|Education        |                   |           |           |
|-----------------+---------------|           |           |
|<HS              |Male   |       | 22625.90| 25060.91|
|                 |---------------+-----------+-----------|
|                 |Female |       | 14850.22| 14449.73|
|-----------------+---------------+-----------+-----------|
|HS               |Male   |       | 31085.59| 28307.08|
|                 |---------------+-----------+-----------|
|                 |Female |       | 18189.30| 17574.26|
|-----------------+---------------+-----------+-----------|
|COLLEGE          |Male   |       | 47960.27| 42817.40|
|                 |---------------+-----------+-----------|
|                 |Female |       | 27745.10| 28767.81|
-------------------------------------------------------------
```

Modifying Row Heading Widths

Have you ever noticed how some tables have lots of room to spare for row labels and others seem very crowded? This is because PROC TABULATE has a basic rule for how to allot space for row headings. Unfortunately, this rule doesn't take the size of the row variable labels into account.

What PROC TABULATE does is look at the LINESIZE setting. The exact formula for the width of the row headings is (LINESIZE/4)-2. If you have a narrow page width with long row variable labels, you end up with row headings that look squashed. Conversely, if you have a wide page width with short row variable labels, you end up with row headings with lots of wasted space, as in the following example:

```
OPTIONS LINESIZE=132;
PROC TABULATE DATA=TEMP;
   CLASS EDUC GENDER OCCUP;
   VAR INCOME;
   TABLE EDUC='Education'*GENDER=' ',
      OCCUP='Occupation'*INCOME=' '*MEAN=' '
      / BOX='Mean Income';
RUN;
```

If you look at Output 10.13, you can see the how the default row heading space looks. The row headings are given a width of 31 or (132/4)-2. Because this table has short row variable labels, this is more space than is necessary. We could always reduce the LINESIZE setting, but this would reduce the space available for the columns. Instead, what we need to do is override the default PROC TABULATE row heading width setting.

We do this with another TABLE statement option called RTS=. The RTS stands for Row Title Space, which is what PROC TABULATE calls the row heading width. You can force PROC TABULATE to allot an exact number of spaces to the row headings by setting RTS= to the width you want. You can figure out the best width by counting up the number of spaces needed to fit your headings. For our example table, RTS=24 allows enough space for the widest labels in the row dimension.

```
PROC TABULATE DATA=TEMP;
   CLASS EDUC GENDER OCCUP;
   VAR INCOME;
   TABLE EDUC='Education'*GENDER=' ',
      OCCUP='Occupation'*INCOME=' '*MEAN=' '
      / BOX='Mean Income' RTS=24;
RUN;
```

As you can see from Output 10.14, the row headings are now only 24 spaces wide, instead of the original 31. All of the labels still fit in this space, and no space is wasted.

Unfortunately, there are limits to the RTS= option. You can set only the overall width of the row headings, you can't control how that space is divided between the variables in the row heading. If you have two nested variables, each will be given half of the available space, even if one variable has an extremely short label and the other variable has an extremely long label.

Output 10.13

```
-----------------------------------------------------------------------------
|Mean Income                     |                 Occupation                |
|                                |-------------------------------------------|
|                                | Managerial |Professional| Technical |  Sales    |
|-------------------------------+------------+------------+-----------+-----------|
|Education         |             |            |            |           |           |
|-----------------+-------------|            |            |           |           |
|<HS              |Male         |   21430.80 |  30075.00  |         . |  28142.33 |
|                 |-------------+------------+------------+-----------+-----------|
|                 |Female       |   31341.00 |  12874.00  |  19615.00 |  10569.64 |
|-----------------+-------------+------------+------------+-----------+-----------|
|HS               |Male         |   41699.72 |  36135.83  |  36886.80 |  35163.25 |
|                 |-------------+------------+------------+-----------+-----------|
|                 |Female       |   24307.68 |  25135.55  |  24048.46 |  13449.10 |
|-----------------+-------------+------------+------------+-----------+-----------|
|College          |Male         |   59390.75 |  50543.66  |  34089.30 |  47632.46 |
|                 |-------------+------------+------------+-----------+-----------|
|                 |Female       |   34018.64 |  32809.55  |  25778.68 |  24876.60 |
-----------------------------------------------------------------------------
```

Output 10.14

```
--------------------------------------------------------------------------
|Mean Income             |                 Occupation                     |
|                        |------------------------------------------------|
|                        | Managerial |Professional| Technical |  Sales    |
|-----------------------+------------+------------+-----------+-----------|
|Education  |           |            |            |           |           |
|----------+-----------|            |            |           |           |
|<HS       |Male        |   21430.80 |  30075.00  |         . |  28142.33 |
|          |-----------+------------+------------+-----------+-----------|
|          |Female      |   31341.00 |  12874.00  |  19615.00 |  10569.64 |
|----------+-----------+------------+------------+-----------+-----------|
|HS        |Male        |   41699.72 |  36135.83  |  36886.80 |  35163.25 |
|          |-----------+------------+------------+-----------+-----------|
|          |Female      |   24307.68 |  25135.55  |  24048.46 |  13449.10 |
|----------+-----------+------------+------------+-----------+-----------|
|College   |Male        |   59390.75 |  50543.66  |  34089.30 |  47632.46 |
|          |-----------+------------+------------+-----------+-----------|
|          |Female      |   34018.64 |  32809.55  |  25778.68 |  24876.60 |
--------------------------------------------------------------------------
```

Modifying Row Heading Indents

There's another way you can reduce the space taken up by the row headings. By default, PROC TABULATE puts nested row headings side by side. If you have multiple levels of nested row variables, PROC TABULATE gives each variable its own "column" in the row heading.

In the following example table, there are three nested row variables. Although the row headings have been reduced as much as possible by using the RTS= option, they still take up a lot of room.

```
PROC TABULATE DATA=TEMP;
   CLASS EDUC GENDER RACE OCCUP;
   VAR INCOME;
   TABLE EDUC='Education'*GENDER=' '*RACE=' ',
      OCCUP='Occupation'*INCOME=' '*MEAN=' '
      / BOX='Mean Income' RTS=32;
RUN;
```

In Output 10.15, you can see that the three row variables are listed side-by-side. We can't make the row headings any narrower by using the RTS= option because the label "Education" won't fit in a space any smaller.

The way around this is to use the INDENT= option.[1] This option does two things. First, it puts the row variable labels on top of each other, instead of side-by-side. Second, it uses the value you specify to indent each of the nested variables a set number of spaces from the variable listed above. The following code shows that by using INDENT=5 for our example table, we can reduce the row title space setting from RTS=32 to RTS=20 and still have plenty of room for the row labels:

```
TABLE EDUC='Education'*GENDER=' '*RACE=' ',
   OCCUP='Occupation'*INCOME=' '*MEAN=' '
   / BOX='Mean Income' RTS=20 INDENT=5;
```

In Output 10.16, you can see that the row labels do indeed fit in a row heading only 20 spaces wide. However, to gain this space, there are some losses.

First, CLASS variable labels are dropped. Note how the label 'Education' is no longer used. If your classification variable values are not clearly understood without their labels, you may have to reformat them or else give up on the INDENT= option. Second, because the row labels are no longer side-by-side, extra rows have to be created to hold the labels. This means that although you can make your table narrower with INDENT= and RTS=, the table is going to get longer. If you can't afford to give up this vertical space, you will not be able to use the INDENT= option.

There is a way to keep your table from growing so long when you use the INDENT= option. Check out "Removing the Row Separators" in Chapter 13.

---

[1]   The INDENT= option was added in Release 6.10.

Output 10.15

Mean Income			Managerial	Professional	Technical	Sales
Education						
HS graduate	Male	White	35574.48	35514.64	29200.47	25717.42
		Black	21000.00	7536.00	26902.17	18498.00
	Female	White	21026.63	20131.81	20561.65	13475.94
		Black	15440.00	31028.00	.	13307.44
College	Male	White	55013.96	51090.75	35060.45	39876.40
		Black	41746.77	53840.20	23898.80	21820.71
	Female	White	27574.87	32154.79	24090.49	19788.56
		Black	21955.64	27297.13	21025.00	22112.29

Output 10.16

Mean Income	Managerial	Professional	Technical	Sales
HS graduate				
Male				
White	35574.48	35514.64	29200.47	25717.42
Black	21000.00	7536.00	26902.17	18498.00
Female				
White	21026.63	20131.81	20561.65	13475.94
Black	15440.00	31028.00	.	13307.44
College				
Male				
White	55013.96	51090.75	35060.45	39876.40
Black	41746.77	53840.20	23898.80	21820.71
Female				
White	27574.87	32154.79	24090.49	19788.56
Black	21955.64	27297.13	21025.00	22112.29

Modifying Column Widths

Now that we've got the row headings under control, let's look at the column headings. Just as PROC TABULATE has a rule for how to pick the width of the row headings, it also has a rule for the width of the column headings. Each column is as wide as the *format* of the variable it contains. All of the tables you've seen so far have had a column width of 12 because the default format for table values is BEST12.2. This is set as the default because it will work for a wide variety of values, but it's not always the format you would pick if you had the choice.

Luckily, PROC TABULATE does give you the choice. You can set the format to be used for your table by specifying the FORMAT= option in the PROC TABULATE statement. For example, the following code specifies a format only 8 spaces wide with no decimal places instead of the default BEST12.2 in:

```
PROC TABULATE DATA=TEMP FORMAT=8.;
   CLASS EDUC RACE FULLTIME;
   VAR INCOME;
   TABLE EDUC='Education'*RACE=' ',
   FULLTIME=' '*INCOME=' '*(N MEAN) / BOX='Income';
RUN;
```

The results are shown in Output 10.17. This width is large enough to hold the table values, but it saves space compared to the default width of 12 spaces. You can also use formats to make larger numbers fit in a table. For example, in this next table, we show sums instead of means.

```
PROC TABULATE DATA=TEMP FORMAT=COMMA12.;
   CLASS EDUC RACE FULLTIME;
   VAR INCOME;
   TABLE EDUC='Education'*RACE=' ',
      FULLTIME=' '*INCOME=' '*(N SUM) / BOX='Income';
RUN;
```

The results are shown in Output 10.18. Now each column is again 12 spaces wide, the default width, but there is room for larger numbers because there are no decimal places displayed as there are with the default BEST12.2 format. Also, the commas make these numbers easier to read.

You will probably find that you need to experiment with your table FORMAT settings. You need to use formats wide enough to leave room for your column headings and wide enough to hold the table values, but not so wide that the table will not fit on the page. The following example shows how you can squeeze a lot of information into a very small space if you absolutely have to. The resulting table is shown in Output 10.19.

```
PROC TABULATE DATA=TEMP FORMAT=7.4;
   CLASS EDUC RACE FULLTIME;
   VAR RATIO;
   TABLE EDUC=' 'Education'*RACE=' ',
      FULLTIME=' '* RATIO =' '*(MEAN STD)
      / BOX='Ratio' RTS=22;
RUN;
```

Output 10.17

		Employed Full-Time		Employed Part-Time	
		N	MEAN	N	MEAN
Education					
HS	White	1391	23962	197	22729
	Black	30	17984	6	15503
COLLEGE	White	2131	38364	257	35195
	Black	54	34382	5	23921

Output 10.18

		Employed Full-Time		Employed Part-Time	
		N	SUM	N	SUM
Education					
HS	White	1,391	33,331,675	197	4,477,628
	Black	30	539,517	6	93,018
COLLEGE	White	2,131	81,753,774	257	9,045,093
	Black	54	1,856,616	5	119,605

Output 10.19

		Employed Full-Time		Employed Part-Time	
		MEAN	STD	MEAN	STD
Education					
HS	White	0.9351	0.2557	0.9350	0.2510
	Black	1.2544	0.2494	1.0687	0.1836
COLLEGE	White	0.9546	0.2531	0.9513	0.3123
	Black	1.2429	0.2513	1.1242	0.5734

Ordering the Headings

So far we've relabeled headings, deleted headings, and changed the width of headings. But what if you've got the right headings but you don't like the order in which they appear? You can change the order of two variables by reversing their order in the TABLE statement. But if your headings represent the values of a CLASS variable, you can't list them specifically in the TABLE statement.

To control the order in which the values of a CLASS variable appear in the row or column headings, you need to use the ORDER= option. This option is used in the PROC TABULATE statement. The four settings for this option, INTERNAL, DATA, FORMATTED, and FREQ, are shown in the following examples.

The default value is ORDER=INTERNAL, and that is what you have been seeing in all of the previous examples. ORDER=INTERNAL means that the class values will be ordered based on their unformatted values. This is basically the same order that you would get if you sorted the data set by the values of the CLASS variable. In the following example, the variable SAT has the values of 1, 2, 3, 4, and 5.

```
PROC TABULATE DATA=TEMP ORDER=INTERNAL;
   CLASS GENDER SAT;
   TABLE GENDER=' ', SAT*N / RTS=15;
RUN;
```

Using ORDER=INTERNAL, we get the table shown in Output 10.20. Notice how the variables are ordered by their values 1, 2, 3, 4, and 5.

The next setting is ORDER=DATA. This setting uses the values in the order in which they appear in the data set. So the order of the variables in your table reflect the sort order of your data set.

```
PROC TABULATE DATA=TEMP ORDER=DATA;
```

In Output 10.21, you can see that this data set must be sorted by descending values of SAT because that's how the variable is ordered in the table. If the data set were not sorted by SAT, the headings would appear in the order in which they are encountered in the data set.

The next setting is ORDER=FORMATTED. This setting uses the formatted values of the variable to order the row headings.

```
PROC TABULATE DATA=TEMP ORDER=FORMATTED;
```

This means that if you have used a format to give the values text labels, they're going to come out in alphabetical order, as in Output 10.22. As you can see, this order makes no sense. However, there are situations where this order can come in handy, as you will see in the next example in this chapter.

The final setting is ORDER=FREQ. This option displays the CLASS variable values in descending order of frequency. In other words, the values with the most observations come first.

```
PROC TABULATE DATA=TEMP ORDER=FREQ;
```

You can see this option in Output 10.23. (The ALL statistic was added to this table to make the ordering clearer.)

Output 10.20

```
-----------------------------------------------------------------------
|             |                     Satisfaction Rating                |
|             |-------------------------------------------------------- |
|             |  POOR=1   |  FAIR=2   |  GOOD=3   |VERY GOOD=4 |EXCELLENT=5 |
|             |-----------+-----------+-----------+-----------+----------- |
|             |     N     |     N     |     N     |     N     |     N      |
|-------------+-----------+-----------+-----------+-----------+----------- |
|Male         |    565.00|    575.00|    558.00|    544.00|    542.00 |
|-------------+-----------+-----------+-----------+-----------+----------- |
|Female       |    626.00|    677.00|    665.00|    732.00|    700.00 |
 -----------------------------------------------------------------------
```

Output 10.21

```
-----------------------------------------------------------------------
|             |                     Satisfaction Rating                |
|             |-------------------------------------------------------- |
|             |EXCELLENT=5 |VERY GOOD=4 |  GOOD=3   |  FAIR=2   |  POOR=1   |
|             |-----------+-----------+-----------+-----------+----------- |
|             |     N     |     N     |     N     |     N     |     N      |
|-------------+-----------+-----------+-----------+-----------+----------- |
|Female       |    700.00|    732.00|    665.00|    677.00|    626.00 |
|-------------+-----------+-----------+-----------+-----------+----------- |
|Male         |    542.00|    544.00|    558.00|    575.00|    565.00 |
 -----------------------------------------------------------------------
```

Output 10.22

```
-----------------------------------------------------------------------
|             |                     Satisfaction Rating                |
|             |-------------------------------------------------------- |
|             |EXCELLENT=5 |  FAIR=2   |  GOOD=3   |  POOR=1   |VERY GOOD=4 |
|             |-----------+-----------+-----------+-----------+----------- |
|             |     N     |     N     |     N     |     N     |     N      |
|-------------+-----------+-----------+-----------+-----------+----------- |
|Female       |    700.00|    677.00|    665.00|    626.00|    732.00 |
|-------------+-----------+-----------+-----------+-----------+----------- |
|Male         |    542.00|    575.00|    558.00|    565.00|    544.00 |
 -----------------------------------------------------------------------
```

Output 10.23

```
-----------------------------------------------------------------------
|             |                     Satisfaction Rating                |
|             |-------------------------------------------------------- |
|             |VERY GOOD=4 |  FAIR=2   |EXCELLENT=5 |  GOOD=3   |  POOR=1   |
|             |-----------+-----------+-----------+-----------+----------- |
|             |     N     |     N     |     N     |     N     |     N      |
|-------------+-----------+-----------+-----------+-----------+----------- |
|Female       |    732.00|    677.00|    700.00|    665.00|    626.00 |
|-------------+-----------+-----------+-----------+-----------+----------- |
|Male         |    544.00|    575.00|    542.00|    558.00|    565.00 |
|-------------+-----------+-----------+-----------+-----------+----------- |
|ALL          |   1276.00|   1252.00|   1242.00|   1223.00|   1191.00 |
 -----------------------------------------------------------------------
```

Reordering the Headings

The previous example hinted that there was a good reason to use ORDER=FORMATTED. The reason is that it lets you force a particular order in your CLASS variable values. If you have a CLASS variable where the values can not be sorted into the correct order, you can use PROC FORMAT to give them new values that can be sorted into the right order. For example, let's say we want to build a table showing test scores from three courses: Reading, Writing, and Arithmetic, in exactly that order. Here is the code to produce the basic table of scores.

```
PROC FORMAT;
   VALUE COURSEFT    101='Writing'
                     104='Arithmetic'
                     109='Reading';
RUN;
PROC TABULATE DATA=TEMP;
   CLASS COURSE GENDER;
   FORMAT COURSE COURSEFT.;
   VAR SCORE;
   TABLE COURSE*MEAN=' ', (GENDER=' ' ALL)*SCORE='Mean Score'
      / ROW=FLOAT;
RUN;
```

However, if you look at Output 10.24, you can see that PROC TABULATE did not put the headings in the order we wanted. Instead of Reading, Writing, and then Arithmetic, we got Writing, Arithmetic, then Reading. This is because the default ORDER setting is ORDER=INTERNAL, which means that SAS orders the values based on their unformatted values (in other words 101, then 104, then 109). So instead, we are going to use ORDER=FORMATTED to get PROC TABULATE to use the formatted values of COURSE, instead of the raw numbers.

```
PROC TABULATE DATA=TEMP ORDER=FORMATTED;
```

As you can see from Output 10.25, this isn't the whole solution. When you sort the formatted values in alphabetical order, you get Arithmetic, then Reading, then Writing. To solve this problem, we need to trick PROC TABULATE into sorting the variables in the correct order. We can do this by putting leading blanks in the format labels. PROC TABULATE will use the blanks when sorting the values into order, but it will drop the blanks before printing the table. To get the values set up correctly, we need to precede our PROC TABULATE code with this revised PROC FORMAT. The value we want to see first has two leading blanks, the value we want to see second has one leading blank, and the value we want to see third has no leading blanks.

```
PROC FORMAT;
   VALUE COURSEFT    109='  Reading'
                     101=' Writing'
                     104='Arithmetic';
RUN;
```

The results are shown in Output 10.26. Notice how the leading blanks have been removed, and the variables are now in the correct order. This is a handy trick to use whenever you want to reorder your headings.

Output 10.24

	Male	Female	ALL
	Mean Score	Mean Score	Mean Score
Course			
Writing	74.50	75.50	74.92
Arithmetic	74.78	75.74	75.30
Reading	72.31	75.14	73.74

Output 10.25

	Female	Male	ALL
	Mean Score	Mean Score	Mean Score
COURSE			
Arithmetic	75.74	74.78	75.30
Reading	75.14	72.31	73.74
Writing	75.50	74.50	74.92

Output 10.26

	Female	Male	ALL
	Mean Score	Mean Score	Mean Score
Course			
Reading	75.14	72.31	73.74
Writing	75.50	74.50	74.92
Arithmetic	75.74	74.78	75.30

Version 7 gives you even more control over the order of the CLASS values in your table. You can use multiple CLASS statements with different ORDER= options to give each classification variable its own ordering, instead of having to pick one ORDER= setting to use for the entire table. See Chapter 24, "Version 7 Enhancements," for details.

Formatting Table Values

To produce meaningful and easy to read tables, it's important that the values are correctly formatted. For example, percentages should have percent signs, dollar amounts should have dollar signs, and all values should show the appropriate number of decimal places.

The Example

PROC TABULATE gives you the power to control the format of every row or column in your data. This allows you to pick the best format for each type of data in your table.

Instead of a "one size fits all" approach where every value in the table has the same format, assign a variety of formats within the same table so that you can produce a table like this example.

	Education					
	<HS		HS		COLLEGE	
	N	MEAN	N	MEAN	N	MEAN
Age	9999	99.9	9999	99.9	9999	99.9
Income	9999	$99,999	9999	$99,999	9999	$99,999

The chapter also shows some formatting tricks and how to produce a table with weighted results.

Using Appropriate Formats - I

Formatting table values was presented in the previous chapter as a way to control column widths, but the FORMAT option can do much more than that. You can control the formatting of your table row by row or column by column.

But first, a quick review of the basics. If you don't specify a value for the FORMAT= option, PROC TABULATE will use the format BEST12.2. This is a "one size fits all" solution that doesn't work well for tables like the following:

```
PROC TABULATE DATA=TEMP;
   CLASS SECTOR OCCUP;
   VAR INCOME;
   TABLE OCCUP=' ', SECTOR=' '*INCOME*(N MEAN);
RUN;
```

Output 11.1 shows how the BEST12.2 format gives our values of N two decimal places. This is wasted space because N will always be an integer. Also, our values of income are dollars, so why not format them as such.

The following code gets rid of the decimal places and adds dollar signs to the numbers. We use a FORMAT= *option* in the PROC TABULATE statement to set the format for table values. This is different than the FORMAT *statement* in PROC TABULATE, which controls the appearance of CLASS variable values. The only way to format the values in table cells is via the FORMAT= option.

```
PROC TABULATE DATA=TEMP FORMAT=DOLLAR10.;
   CLASS SECTOR OCCUP;
   VAR INCOME;
   TABLE OCCUP=' ', SECTOR=' '*INCOME*(N MEAN);
RUN;
```

In Output 11.2, you can see that we just made the problem worse. Now we have Ns with dollar signs. Unfortunately, if you specify a FORMAT in the PROC TABULATE statement, it applies to the whole table.

There is a way to get around this. The FORMAT= option is also valid in the TABLE statement, and you can apply it to individual variables or statistics. You do this is with the asterisk operator. To format the N statistic as a six-digit integer, you use the syntax "N*FORMAT=6." Or, to save space, you can abbreviate FORMAT to F and use "N*F=6." Then you can use a different format for the MEAN statistic in the same manner.

```
PROC TABULATE DATA=TEMP;
   CLASS SECTOR OCCUP;
   VAR INCOME;
   TABLE OCCUP=' ',
      SECTOR=' '*INCOME*(N*F=6. MEAN*F=DOLLAR10.);
RUN;
```

If you look at Output 11.3, you can see that the N statistic is now shown as an integer in a column 6 spaces wide, and the mean is shown as a dollar amount in a column 10 spaces wide.

Output 11.1

	Private		Public	
	Income		Income	
	N	MEAN	N	MEAN
Managerial	672.00	46075.54	96.00	44390.04
Professional	632.00	42098.03	233.00	36029.82
Technical	127.00	30209.65	17.00	28168.53
Sales	442.00	31988.68	5.00	50317.80

Output 11.2

	Private		Public	
	Income		Income	
	N	MEAN	N	MEAN
Managerial	$672	$46,076	$96	$44,390
Professional	$632	$42,098	$233	$36,030
Technical	$127	$30,210	$17	$28,169
Sales	$442	$31,989	$5	$50,318

Output 11.3

	Private		Public	
	Income		Income	
	N	MEAN	N	MEAN
Managerial	672	$46,076	96	$44,390
Professional	632	$42,098	233	$36,030
Technical	127	$30,210	17	$28,169
Sales	442	$31,989	5	$50,318

Using Appropriate Formats - II

The previous example used two different formats for two different statistics, but you might also want to use two different formats for two different analysis variables. In the following table, AGE is formatted as a number with a single decimal place, and INCOME is shown in dollars:

```
PROC TABULATE DATA=TEMP;
   CLASS EDUC;
   VAR AGE INCOME;
   TABLE AGE*F=6.1 INCOME*F=DOLLAR10., EDUC*MEAN;
RUN;
```

The resulting table is shown in Output 11.4. Each analysis variable in the table now has an appropriate format. Also, notice how each of the columns is 10 spaces wide. Even though AGE is formatted as 6 spaces wide, PROC TABULATE can't produce a column with two widths, so both widths are set to 10.

But what do you do if you have two analysis variables that each require a particular format, and you also have two statistics that each require a particular format? For example, what if you want to show both the N and the MEAN in the table above. You could format the age as a number and the income in dollars:

```
TABLE AGE*F=6.1 INCOME*F=DOLLAR10., EDUC*(N MEAN);
```

Or, you could format the N as an integer and the mean with a decimal place:

```
TABLE AGE INCOME, EDUC*(N*F=8. MEAN*F=8.1);
```

These two tables are shown in Output 11.5 and Output 11.6. Each table gets part of the formatting right. But how do you get all of the formatting right at the same time? PROC TABULATE will let you select formats for more than one dimension of the table. It's a bit tricky, but if you set up the formats in the right order, you can have statistics that vary both by row and by column. The following code creates formats that vary both by variable and by statistic:

```
TABLE AGE*F=6.1 INCOME*F=DOLLAR10.,
   EDUC*(N*F=6. MEAN);
```

In the table shown in Output 11.7, there are three different formats applied to table values. The N statistic is formatted as an integer, the means for AGE are formatted with one decimal place, and the means for INCOME are formatted as dollars.

The way this works is that PROC TABULATE evaluates the formats in the order that they are listed. First AGE is formatted as 6.1, then INCOME is formatted as DOLLAR10., then comes the 6. format for the N statistic. This 6. format overrides the 6.1 format for the N of AGE, and the DOLLAR10. format for the N of INCOME. Because the MEAN is not given a format, the MEAN of AGE remains formatted as 6.1 and the MEAN of INCOME remains formatted as DOLLAR10.

This type of formatting will not work in all situations. Sometimes your desired formats are just too complex, and you will have to revise the arrangement of the variables in the table in order to format them correctly.

Output 11.4

```
-----------------------------------------------------------------
|                              |            EDUCATION            |
|                              |---------------------------------|
|                              |  <HS   |   HS    | COLLEGE      |
|                              |----------+----------+---------- |
|                              |  MEAN  |  MEAN   |  MEAN        |
|------------------------------+----------+----------+---------- |
|AGE                           |   59.5 |    49.5 |    44.8      |
|------------------------------+----------+----------+---------- |
|INCOME                        | $11,938 | $19,582 | $32,711     |
-----------------------------------------------------------------
```

Output 11.5

```
-----------------------------------------------------------------------------------------
|                   |                            EDUCATION                               |
|                   |-------------------------------------------------------------------|
|                   |      <HS        |        HS          |        COLLEGE             |
|                   |-------------------+--------------------+-------------------------- |
|                   | N    |  MEAN    |  N     |  MEAN     |  N      |  MEAN             |
|-------------------+----------+---------+----------+---------+----------+---------------|
|AGE                | 1001.0 |   59.5 | 2282.0 |   49.5 | 3356.0 |   44.8               |
|-------------------+----------+---------+----------+---------+----------+---------------|
|INCOME             | $1,001 | $11,938 | $2,282 | $19,582 | $3,356 | $32,711            |
-----------------------------------------------------------------------------------------
```

Output 11.6

```
-------------------------------------------------------------------------------
|                           |                    Education                    |
|                           |-------------------------------------------------|
|                           |   <HS      |   HS graduate   |    College        |
|                           |---------------+---------------+---------------- |
|                           | N  |  MEAN  |  N  |  MEAN   |  N  |  MEAN         |
|---------------------------+--------+--------+--------+--------+--------+----- |
|Age                        | 1001 |   59.5 | 2282 |   49.5 | 3356 |   44.8    |
|---------------------------+--------+--------+--------+--------+--------+----- |
|Income                     | 1001 | 11937.5 | 2282 | 19581.8 | 3356 | 32710.6 |
-------------------------------------------------------------------------------
```

Output 11.7

```
-------------------------------------------------------------------------------
|                           |                    EDUCATION                    |
|                           |-------------------------------------------------|
|                           |   <HS      |      HS       |    COLLEGE          |
|                           |---------------+---------------+---------------- |
|                           | N  |  MEAN  |  N  |  MEAN   |  N  |  MEAN         |
|---------------------------+------+---------+------+---------+------+--------- |
|AGE                        | 1001 |   59.5 | 2282 |   49.5 | 3356 |   44.8    |
|---------------------------+------+---------+------+---------+------+--------- |
|INCOME                     | 1001 | $11,938 | 2282 | $19,582 | 3356 | $32,711 |
-------------------------------------------------------------------------------
```

Formatting Percentages

You may have noticed back in Chapters 7 and 8 that when PROC TABULATE computes percentages, it does not include percent signs in the results. If we use row or column headings that clearly label the results as percentages, this is okay. But what about when we want to get rid of excess headings, as in the following table?

```
PROC TABULATE DATA=TEMP;
   CLASS OCCUP GENDER;
   TABLE OCCUP*PCTN=" ", GENDER ALL / ROW=FLOAT;
RUN;
```

In Output 11.8, you can see that the table is not very clear. The values displayed are percentages, but there's no easy way to see that from the labeling of the table. What we need to do is use the FORMAT option to apply an appropriate format. The following example uses the PERCENT9.1 format to add percent signs to the values.

```
PROC TABULATE DATA=TEMP;
   CLASS OCCUP GENDER;
   TABLE OCCUP*PCTN=" "*F=PERCENT9.1, GENDER ALL / ROW=FLOAT;
RUN;
```

But if you look at Output 11.9, you can see that something has gone wrong. The values are much too large. This is because the standard PERCENT format first multiplies values by 100 and then adds the percent sign. But PROC TABULATE has already multiplied the values by 100, so what happens is that the values get multiplied by 1000!

To get around this, we need to create our own format for percentages. You can do this with a PICTURE format. A picture format is a description of how you would like the values to be displayed. In the following code, the zeros are used to indicate the positions where we want to see a number. If the value is not as wide as the format, the zeros tell PROC TABULATE to leave the leading spaces blank. The decimal place indicates that we want our percentages to have a single decimal place. The percent sign is then added to the end of each number.

You apply a PICTURE format to the table exactly the same way you apply any other format, by using the FORMAT option. In this case, we use F=PCTPIC. to select our percentage format.

```
PROC FORMAT;
   PICTURE PCTPIC LOW-HIGH='000.0%';
RUN;
```

```
PROC TABULATE DATA=TEMP;
   CLASS OCCUP GENDER;
   TABLE OCCUP*PCTN=" "*F=PCTPIC., GENDER ALL / ROW=FLOAT;
RUN;
```

The results are shown in Output 11.10. Now each percentage has a percent sign, but the values have not been multiplied by 100. This PICTURE format is a handy tool to use with any table that is displaying percentages.

Output 11.8

```
-----------------------------------------------------------------
|                             |           Gender       |          |
|                             |------------------------|          |
|                             | Female   |   Male      |   ALL    |
|-----------------------------+----------+-------------+--------- |
|Occupation                   |          |             |          |
|-----------------------------|          |             |          |
|Managerial                   |    19.90 |      14.69  |   34.59  |
|-----------------------------+----------+-------------+--------- |
|Professional                 |    17.79 |      21.07  |   38.86  |
|-----------------------------+----------+-------------+--------- |
|Technical                    |     3.19 |       3.28  |    6.47  |
|-----------------------------+----------+-------------+--------- |
|Sales                        |    11.05 |       9.03  |   20.08  |
-----------------------------------------------------------------
```

Output 11.9

```
------------------------------------------------------------
|                             |        Gender    |          |
|                             |------------------|          |
|                             | Female |  Male   |   ALL    |
|-----------------------------+--------+---------+--------- |
|Occupation                   |        |         |          |
|-----------------------------|        |         |          |
|Managerial                   | 1990.1%| 1469.0% | 3459.1%  |
|-----------------------------+--------+---------+--------- |
|Professional                 | 1779.0%| 2106.9% | 3885.9%  |
|-----------------------------+--------+---------+--------- |
|Technical                    |  319.0%|  327.9% |  646.9%  |
|-----------------------------+--------+---------+--------- |
|Sales                        | 1105.1%|  903.0% | 2008.1%  |
------------------------------------------------------------
```

Output 11.10

```
-----------------------------------------------------
|                            |  Gender       |        |
|                            |-------------- |        |
|                            |Female| Male | ALL   |
|----------------------------+------+------+------ |
|Occupation                  |      |      |        |
|----------------------------|      |      |        |
|Managerial                  | 19.9%| 14.6%| 34.5% |
|----------------------------+------+------+------ |
|Professional                | 17.7%| 21.0%| 38.8% |
|----------------------------+------+------+------ |
|Technical                   |  3.1%|  3.2%|  6.4% |
|----------------------------+------+------+------ |
|Sales                       | 11.0%|  9.0%| 20.0% |
-----------------------------------------------------
```

Formatting Large Numbers

The goal in using formats is to display the values in your table so that they are easy to view and understand. When you are dealing with large numbers, this can be a challenge.

The following code produces a table with some large dollar values:

```
PROC TABULATE DATA=TEMPBIG;
   CLASS SECTOR OCCUP;
   VAR INCOME;
   TABLE SECTOR, OCCUP*INCOME=' '*SUM=' '
      / BOX='Total Income' RTS=15;
RUN;
```

Because we did not specify a format, the default BEST12.2 was used. In the resulting table, shown in Output 11.11, we have a mess. Some of the numbers have two decimal places, some have one, and some have none. All of the numbers are very hard to read.

Because these values are dollars, we can switch to a more appropriate DOLLARw.d format. To have room for the largest numbers, we'll drop the decimal places, and allow the numbers to be up to 15 digits by using a DOLLAR15. format.

```
PROC TABULATE DATA=TEMPBIG;
   CLASS SECTOR OCCUP;
   VAR INCOME;
   TABLE SECTOR, OCCUP*INCOME=' '*SUM=' '*F=DOLLAR15.
      / BOX='Total Income' RTS=15;
RUN;
```

The second version of our table, shown in Output 11.12, looks much better. At least now all of the table cells have the same format. However, these numbers are so large that the table is still hard to read. When you have extremely large numbers, you should consider getting rid of some of the digits. Instead of reporting total income in dollars, we can report total income in millions of dollars.

To do this, we use a PROC FORMAT to set up a picture format, and then use the MULT option to convert dollars to millions of dollars. The following code produces numbers formatted with dollar signs and commas and converted into millions. The resulting table is shown in Output 11.13.

```
PROC FORMAT;
   PICTURE MILPIC LOW-HIGH='000,000' (PREFIX='$' MULT=.000001);
RUN;
PROC TABULATE DATA=TEMPBIG;
   CLASS SECTOR OCCUP;
   VAR INCOME;
   TABLE SECTOR, OCCUP*INCOME=' '*SUM=' '*F=MILPIC12.
      / BOX='Total Income (in Millions)' RTS=15;
RUN;
```

Output 11.11

```
-----------------------------------------------------------------
|Total Income |                  Occupation                     |
|             |------------------------------------------------ |
|             | Managerial |Professional| Technical |  Sales    |
|-------------+------------+------------+-----------+---------- |
|Sector       |            |            |           |           |
|-------------|            |            |           |           |
|Private      | 32232236283| 27696802278|3993926625.0| 14718695877 |
|-------------+------------+------------+-----------+---------- |
|Public       |4436163204.0|8739140868.0|498498465.00|261904149.00 |
-----------------------------------------------------------------
```

Output 11.12

```
--------------------------------------------------------------------------
|Total Income |                      Occupation                           |
|             |-------------------------------------------------------- |
|             | Managerial  | Professional  |  Technical   |   Sales      |
|-------------+-------------+---------------+--------------+------------- |
|Sector       |             |               |              |              |
|-------------|             |               |              |              |
|Private      |$32,232,236,283|$27,696,802,278| $3,993,926,625|$14,718,695,877 |
|-------------+-------------+---------------+--------------+------------- |
|Public       | $4,436,163,204| $8,739,140,868|   $498,498,465|   $261,904,149 |
--------------------------------------------------------------------------
```

Output 11.13

```
-----------------------------------------------------------------
|Total Income |                  Occupation                     |
|(in Millions)|------------------------------------------------ |
|             | Managerial |Professional| Technical |  Sales    |
|-------------+------------+------------+-----------+---------- |
|Sector       |            |            |           |           |
|-------------|            |            |           |           |
|Private      |    $32,232 |    $27,696 |    $3,993 |   $14,718 |
|-------------+------------+------------+-----------+---------- |
|Public       |     $4,436 |     $8,739 |      $498 |     $261  |
-----------------------------------------------------------------
```

Weighting the Results

There is one other way that you can alter the values in a table, and that is by weighting them. This isn't really a reformatting of the table values but rather a recalculating of the table values.

Weighting is frequently used when the data represent a sample of a larger population. For example, you might have data from an opinion poll of 1000 adults nationwide. To extrapolate the results to the national population, you use weights. For example, if the national population is 50% female and 50% male, but your sample is only 40% female, you can to use weights to adjust your results. If you give each observation for females a weight of 1.25 (.50/.40) and each observation for males a weight of 0.83 (.50/.60), you can adjust for this difference in gender distribution.

The following code shows how a table would look for our 1000 observations if we do not use any weights:

```
PROC TABULATE DATA=TEMP F=6.;
   CLASS YEAR GENDER;
   VAR SAT;
   TABLE (YEAR=" " ALL),
      (GENDER=" " ALL)*SAT=" "*(N*F=6. MEAN*F=6.1)
      / RTS=20 BOX="Satisfaction Rating" ROW=FLOAT;
RUN;
```

In Output 11.14, you can see that our table has a total of 400 observations for females and 600 observations for males. This is an unweighted table and is not representative of the national population. To adjust for the gender imbalance in our data, we need to apply our weights. To produce a weighted table in PROC TABULATE, you use the WEIGHT statement. The syntax is "WEIGHT *varname*", where *varname* is the name of the variable that holds the weight for each observation.

The following code produces a weighted table. There are two changes in this code. First, the WEIGHT statement is added. Second, the N statistic is replaced by the SUMWGT statistic. This is because N statistics are always unweighted, whether or not you use a WEIGHT statement. The SUMWGT statistic adds up the values of each of the weights to create a weighted number of observations. For this reason, SUMWGT can be labeled as "N" in this example.

```
PROC TABULATE DATA=TEMP;
   CLASS YEAR GENDER;
   VAR SAT;
   WEIGHT THEWT;
   TABLE (YEAR=" " ALL),
      (GENDER=" " ALL)*SAT=" "*(SUMWGT="N"*F=6. MEAN*F=6.1)
      / RTS=20 BOX="Weighted Satisfaction Rating" ROW=FLOAT;
RUN;
```

The resulting weighted table is shown in Output 11.15. Notice how the observations now add up to 500 females and 500 males. And the means have been adjusted, too. The means for females and males do not change because the increase in observations does not affect the mean, but the overall means do change. This is because females had lower satisfaction ratings then males, so if you increase the percentage of females from 40% to 50%, the mean overall satisfaction ratings will go down.

Output 11.14

Satisfaction Rating		Male		Female		ALL	
		N	MEAN	N	MEAN	N	MEAN
1990		107	5.1	59	3.3	166	4.4
1991		91	4.8	65	2.7	156	3.9
1992		44	5.2	27	2.8	71	4.3
1993		50	5.0	42	2.8	92	4.0
1994		44	4.8	20	3.0	64	4.2
1995		11	4.8	9	2.9	20	4.0
1996		19	5.2	13	3.5	32	4.5
1997		7	4.0	6	2.2	13	3.2
1998		227	4.9	159	3.1	386	4.2
ALL		600	5.0	400	3.0	1000	4.2

Output 11.15

Weighted Satisfaction Rating		Male		Female		ALL	
		N	MEAN	N	MEAN	N	MEAN
1990		89	5.1	74	3.3	163	4.3
1991		76	4.8	81	2.7	157	3.7
1992		37	5.2	34	2.8	70	4.1
1993		42	5.0	53	2.8	94	3.8
1994		37	4.8	25	3.0	62	4.1
1995		9	4.8	11	2.9	20	3.8
1996		16	5.2	16	3.5	32	4.4
1997		6	4.0	8	2.2	13	3.0
1998		189	4.9	199	3.1	388	4.0
ALL		500	5.0	500	3.0	1000	4.0

Modifying the Table Grid

Each of the previous chapters has addressed how to vary the contents of a PROC TABULATE table. This chapter looks at the table itself.

If you don't like the standard grid of boxes that PROC TABULATE creates, there are ways you can change it to make the table more aesthetically pleasing.

The Example

In this chapter, we're going to take a basic table and clean it up. We'll change the justification, remove some of the table grid, and create smooth lines for the borders.

Income	Employed PT		Employed FT		Total	
	N	Mean	N	Mean	N	Mean
<HS						
White	999	$99,999	999	$99,999	999	$99,999
Black	999	$99,999	999	$99,999	999	$99,999
Other	999	$99,999	999	$99,999	999	$99,999
HS						
White	999	$99,999	999	$99,999	999	$99,999
Black	999	$99,999	999	$99,999	999	$99,999
Other	999	$99,999	999	$99,999	999	$99,999
College						
White	999	$99,999	999	$99,999	999	$99,999
Black	999	$99,999	999	$99,999	999	$99,999
Other	999	$99,999	999	$99,999	999	$99,999

Left-Justifying the Table

All of the tables that have been displayed so far were centered on the page, but there's no reason tables have to be built this way. In fact, in some circumstances it's better if tables are not centered.

You can control the justification of the table on the page via the SAS system option CENTER|NOCENTER. The default setting is CENTER, which produces tables that are centered in the middle of the page. The middle of the page is based on your current LINESIZE setting. The following code shows a table with the default setting of CENTER:

```
OPTIONS CENTER;
PROC TABULATE DATA=TEMP;
   CLASS OCCUP GENDER;
   VAR AGE;
   TABLE OCCUP=' ', GENDER=' '*AGE=' '*MEAN=' '
      / BOX='Mean Age' RTS=15 ROW=FLOAT;
   TITLE 'THIS IS A CENTERED TITLE AND TABLE';
RUN;
```

The resulting table is shown in Output 12.1. The table is centered in the middle of the page. If you are going to display your tables beneath centered titles or figure labels, as is shown here, then it makes sense to center the table as well. On the other hand, if you prefer your output to be aligned flush left, then you will want to use the other option setting: NOCENTER.

```
OPTIONS NOCENTER;
PROC TABULATE DATA=TEMP;
   CLASS OCCUP GENDER;
   VAR AGE;
   TABLE OCCUP=' ', GENDER=' '*AGE=' '*MEAN=' '
      / BOX='Mean Age' RTS=15 ROW=FLOAT;
   TITLE 'THIS IS A LEFT-JUSTIFIED TITLE AND TABLE';
RUN;
```

The resulting table is shown in Output 12.2. The table is now aligned on the left side of the page.

There is another reason why you might want to left-justify your PROC TABULATE output. If you plan to export your output to another software package such as a word processor or spreadsheet for presentation, then you should always use NOCENTER. The reason for this is that SAS uses leading blanks to center its output. For each row in the table shown in Output 12.1, there are actually dozens of blank spaces preceding the table itself. These leading spaces will confuse a word processor or spreadsheet when you try to import your table. It is better to leave the output left-justified in SAS and then change the justification after the table has been imported into the word processor or spreadsheet. (See Chapter 16, "Outputting the Results," for information on how to export PROC TABULATE output to other software packages.)

Output 12.1

```
            THIS IS A CENTERED TITLE AND TABLE
     -------------------------------------------------
     |Mean Age      |   Female   |    Male    |
     |--------------+------------+------------|
     |Managerial    |      44.67|       41.53 |
     |--------------+------------+------------|
     |Professional  |      41.25|       41.71 |
     |--------------+------------+------------|
     |Technical     |      38.72|       38.58 |
     |--------------+------------+------------|
     |Sales         |      43.15|       42.60 |
     |--------------+------------+------------|
     |Clerical      |      41.01|       42.34 |
     |--------------+------------+------------|
     |Services      |      41.07|       41.86 |
     |--------------+------------+------------|
     |Manufacturing |      41.62|       42.53 |
     |--------------+------------+------------|
     |Transport     |      42.44|       40.39 |
     |--------------+------------+------------|
     |Farming       |      41.39|       40.54 |
     -------------------------------------------------
```

Output 12.2

```
THIS IS A LEFT-JUSTIFIED TITLE AND TABLE

-------------------------------------------------
|Mean Age      |   Female   |    Male    |
|--------------+------------+------------|
|Managerial    |      44.67|       41.53|
|--------------+------------+------------|
|Professional  |      41.25|       41.71|
|--------------+------------+------------|
|Technical     |      38.72|       38.58|
|--------------+------------+------------|
|Sales         |      43.15|       42.60|
|--------------+------------+------------|
|Clerical      |      41.01|       42.34|
|--------------+------------+------------|
|Services      |      41.07|       41.86|
|--------------+------------+------------|
|Manufacturing |      41.62|       42.53|
|--------------+------------+------------|
|Transport     |      42.44|       40.39|
|--------------+------------+------------|
|Farming       |      41.39|       40.54|
-------------------------------------------------
```

Removing the Row Dividers

In each of the examples so far, we've used the default PROC TABULATE table grid. The table grid is the dashes, plus signs, and vertical bars that make up the borders of the table rows, columns, and cells. The default table grid puts a border around every cell, row, column, and page.

But you don't have to use the standard table grid. If you'd like a table that looks a little less boxy, you can tell PROC TABULATE to remove some of the dividers. In particular, you can get rid of the row dividers. If you look closely at a PROC TABULATE table, you will see that each row in the table actually requires two lines of text. One has the row heading, row values, and the column dividers. The second line consists of a series of dashes and plus signs that form the divider between the rows.

The following code uses the NOSEPS option to remove these row dividers from your table. By adding the NOSEPS option to your PROC TABULATE statement, you turn these lines off.

```
PROC TABULATE DATA=TEMP NOSEPS;
   CLASS EDUC RACE FULLTIME;
   VAR INCOME;
   TABLE EDUC=" "*(RACE=" " ALL),
      FULLTIME=" "*INCOME=" "*(N*F=6. MEAN*F=DOLLAR8.)
      / BOX="Income";
RUN;
```

Output 12.3 shows what happens when you use NOSEPS. Now each of the rows is stacked right on top of the next row with no dividers. There are no grid lines separating either the row values or the row headings. This can save a lot of space because now your output has only one line of text per row.

Sometimes removing the row separators can also make your table easier to read. On the other hand, if you have a table with a fairly complex structure, removing the separators may make your table harder to read.

One thing you can do to improve readability is to combine the NOSEPS option with the INDENT option. Because removing the separators can make nested row headings hard to read, adding an indent may help. The following code produces the same table as before, but with the addition of a three-space indent:

```
PROC TABULATE DATA=TEMP NOSEPS;
   CLASS EDUC RACE FULLTIME;
   VAR INCOME;
   TABLE EDUC=" "*(RACE=" " ALL="--ALL--"),
      FULLTIME=" "*INCOME=" "*(N*F=6. MEAN*F=DOLLAR8.)
      / BOX="Income" INDENT=3 RTS=12;
RUN;
```

The new table, shown in Output 12.4, is much easier to read. It's also more compact. Using the INDENT option allows the row headings to be reduced to only 12 spaces wide.

Output 12.3

```
--------------------------------------------------------------
|Income                 | Full time     | Part time     |
|                       |---------------+---------------|
|                       | N  | MEAN     | N  | MEAN     |
|-----------------------+------+--------+------+--------|
|<HS         White      | 241| $19,500 |  41| $19,910 |
|            Black       |   9| $15,548 |   2| $14,949 |
|            ALL         | 250| $19,358 |  43| $19,680 |
|HS          White      |1150| $24,897 | 156| $23,470 |
|            Black       |  21| $19,028 |   4| $15,780 |
|            Other       |   2| $25,508 |   1| $13,017 |
|            ALL         |1173| $24,793 | 161| $23,214 |
|College     White      |2131| $38,364 | 257| $35,195 |
|            Black       |  54| $34,382 |   5| $23,921 |
|            Other       |   2|  $7,777 |   .|       . |
|            ALL         |2187| $38,238 | 262| $34,980 |
--------------------------------------------------------------
```

Output 12.4

```
-------------------------------------------------
|Income    | Full time     | Part time     |
|          |---------------+---------------|
|          | N  | MEAN     | N  | MEAN     |
|----------+------+--------+------+--------|
|<HS       |    |          |    |          |
|   White  | 241| $19,500 |  41| $19,910 |
|   Black  |   9| $15,548 |   2| $14,949 |
|  --ALL-- | 250| $19,358 |  43| $19,680 |
|HS        |    |          |    |          |
|   White  |1150| $24,897 | 156| $23,470 |
|   Black  |  21| $19,028 |   4| $15,780 |
|   Other  |   2| $25,508 |   1| $13,017 |
|  --ALL-- |1173| $24,793 | 161| $23,214 |
|College   |    |          |    |          |
|   White  |2131| $38,364 | 257| $35,195 |
|   Black  |  54| $34,382 |   5| $23,921 |
|   Other  |   2|  $7,777 |   .|       . |
|  --ALL-- |2187| $38,238 | 262| $34,980 |
-------------------------------------------------
```

If you look closely at Output 12.4, you'll notice another change that was made to the code. The ALL variable was given a label of "--ALL--" to make it stand out as a subtotal. This is a handy trick to use in tables with NOSEPS turned on.

Modifying the FORMCHAR Setting

The NOSEPS option lets you change one aspect of the table grid. But if you want even more control over the row, column, cell, and table borders, you need to look at the FORMCHAR setting.

Each of the characters used to create the table grid is defined by the FORMCHAR setting. FORMCHAR can be specified as a system option, or you can specify it in the PROC TABULATE statement. The following example shows the default setting of FORMCHAR for all of the tables in the previous examples. It is a string with eleven characters, each of which represents a part of the table grid.

```
PROC TABULATE DATA=TEMP FORMCHAR='|----|+|---';
```

The table in Output 12.5 shows how each of the characters in the FORMCHAR string is used. For example, the first character determines the vertical character used to divide the columns, the second character determines the horizontal character used to divide the rows, and so on.

You can modify the FORMCHAR string to produce a different grid for your table. For example, the following FORMCHAR string produces a table with only vertical dividers. Each of the characters that would have produced a horizontal divider was replaced with a blank in the FORMCHAR string. And the plus sign that is used where row and column dividers intersect was replaced with a vertical bar.

```
PROC TABULATE DATA=TEMP FORMCHAR='|     |||   ';
```

The resulting table is shown in Output 12.6. Notice how the table contents and spacing have not changed at all. The only difference between Output 12.5 and Output 12.6 is that the spaces that used to hold a horizontal character are now blank, and the spaces that used to have plus signs now have vertical bars.

We can also modify the FORMCHAR string to produce a table with only horizontal dividers. In this case, we replaced each of the vertical characters in the FORMCHAR string with either a blank or a horizontal divider.

```
PROC TABULATE DATA=TEMP FORMCHAR='  ----------';
```

The resulting table is shown in Output 12.7. You can compare this to Output 12.5 to see how the changes affected the table layout.

If you are using a more recent release of SAS, it probably has a default FORMCHAR setting like the one shown in the code below. These are special printer characters that will produce a table with smooth connected lines for all borders.

```
PROC TABULATE DATA=TEMP FORMCHAR='|─┌┤├┼';
```

But there's a catch with this setting, as you can see in Output 12.8. These characters are printer specific. The table will look correct in the SAS output window, but when you print, you may get the strange results shown here. This is because the printer couldn't figure out how to display the characters specified in the FORMCHAR string. Chapter 22, "FORMCHAR Settings Reference," shows how you can look up the correct FORMCHAR setting for your printer so that you can generate tables with smooth borders.

Output 12.5

```
------------------------------------------
|Mean Age       |   Female   |   Male    |
|---------------+------------+---------- |
|<HS            |     57.95|    60.69 |
|---------------+------------+---------- |
|HS             |     47.29|    51.40 |
|---------------+------------+---------- |
|College        |     44.79|    44.72 |
------------------------------------------
```

Output 12.6

```
|Mean Age       |   Female   |   Male    |
|               |            |           |
|<HS            |     57.95|    60.69 |
|               |            |           |
|HS             |     47.29|    51.40 |
|               |            |           |
|College        |     44.79|    44.72 |
```

Output 12.7

```
------------------------------------------
    Mean Age        Female      Male
------------------------------------------
   <HS              57.95       60.69
------------------------------------------
   HS               47.29       51.40
------------------------------------------
   College          44.79       44.72
------------------------------------------
```

Output 12.8

```
ÑÉÉÉÉÉÉÉÉÉÉÉÉÉÉÖÉÉÉÉÉÉÉÉÉÉÉÉÉÉÖÉÉÉÉÉÉÉÉÉÉÉÉÉÜ
ÇMean Age       Ç   Female   Ç   Male    Ç
áÉÉÉÉÉÉÉÉÉÉÉÉÉÉÉàÉÉÉÉÉÉÉÉÉÉÉÉÉàÉÉÉÉÉÉÉÉÉÉÉÉÉâ
Ç<HS            Ç     57.95Ç    60.69Ç
áÉÉÉÉÉÉÉÉÉÉÉÉÉÉÉàÉÉÉÉÉÉÉÉÉÉÉÉÉàÉÉÉÉÉÉÉÉÉÉÉÉÉâ
ÇHS             Ç     47.29Ç    51.40Ç
áÉÉÉÉÉÉÉÉÉÉÉÉÉÉÉàÉÉÉÉÉÉÉÉÉÉÉÉÉàÉÉÉÉÉÉÉÉÉÉÉÉÉâ
ÇCollege        Ç     44.79Ç    44.72Ç
áÉÉÉÉÉÉÉÉÉÉÉÉÉÉÉàÉÉÉÉÉÉÉÉÉÉÉÉÉàÉÉÉÉÉÉÉÉÉÉÉÉÉâ
```

Fitting More Tables on the Page: CONDENSE and NOCONTINUED

This example is not about modifying the table grid. Instead, this example looks at the placement of the table itself.

By now you've seen examples of tables that span more than one page. By default, if your table is too wide for one page, PROC TABULATE generates a "(CONTINUED)" label, which it prints beneath the table. Then, a page break is generated and the rest of the table prints on the next page.

If your table is both tall and wide, this is probably the appropriate way to handle things. But if you have a short table that is too wide for the page, you end up wasting lots of paper. A table that is 10 lines long and 2 pages wide is not very efficient. For example, look at the following code.

```
PROC TABULATE DATA=TEMP;
   CLASS GENDER EDUC RACE;
   TABLE RACE, GENDER=' '*EDUC=' '*N=' ' /
      BOX='Number of Observations';
RUN;
```

As you can see in Output 12.9, this table spans two pages (the page break has been removed to save space). But the table is quite short, and both halves of the table would easily fit on one page.

You can tell PROC TABULATE to put more than one table on the page by using the CONDENSE option in the TABLE statement. This option forces PROC TABULATE to put as many tables on the page as will fit. If your table is too tall for more than one section of the table to fit on the page, then CONDENSE has no effect.

If using CONDENSE allows the whole table to fit on one page, you can add another option to get rid of the "(CONTINUED)" label that is now unnecessary. By specifying NOCONTINUED in the TABLE statement, you can get rid of these labels. (If your table still spans multiple pages after you use the CONDENSE option, you probably don't want to use the NOCONTINUED option.)

```
PROC TABULATE DATA=TEMP;
   CLASS GENDER EDUC RACE;
   TABLE RACE, GENDER=' '*EDUC=' '*N=' ' /
      BOX='Number of Observations'
      NOCONTINUED CONDENSE;
RUN;
```

Output 12.10 shows the effect of using both NOCONTINUED and CONDENSE. Now there is no page break between the two halves of the table, and there is no "(CONTINUED)" label.

By the way, all of this only applies to tables that are too *wide* to fit on a page. If your table is too *long* for the page, CONDENSE has no effect, and using NOCONTINUED does not make much sense.

Output 12.9

```
-----------------------------------------------------------------
|Number of       |                   Female                 |   Male    |
|Observations    |------------------------------------------+-----------|
|                |    <HS    |     HS    |   College  |    <HS    |
|----------------+-----------+-----------+------------+-----------|
|Race            |           |           |            |           |
|----------------|           |           |            |           |
|White           |     19.00 |    139.00 |     735.00 |     18.00 |
|----------------+-----------+-----------+------------+-----------|
|Black           |      2.00 |      9.00 |      83.00 |      3.00 |
-----------------------------------------------------------------
```

(CONTINUED)

... *page break deleted to save space* ...

```
---------------------------------------------
|Number of       |         Male         |
|Observations    |----------------------|
|                |     HS    |  College  |
|----------------+-----------+-----------|
|Race            |           |           |
|----------------|           |           |
|White           |    150.00 |    715.00 |
|----------------+-----------+-----------|
|Black           |      9.00 |     53.00 |
---------------------------------------------
```

Output 12.10

```
-----------------------------------------------------------------
|Number of       |                   Female                 |   Male    |
|Observations    |------------------------------------------+-----------|
|                |    <HS    |     HS    |   College  |    <HS    |
|----------------+-----------+-----------+------------+-----------|
|Race            |           |           |            |           |
|----------------|           |           |            |           |
|White           |     19.00 |    139.00 |     735.00 |     18.00 |
|----------------+-----------+-----------+------------+-----------|
|Black           |      2.00 |      9.00 |      83.00 |      3.00 |
-----------------------------------------------------------------
```

```
---------------------------------------------
|Number of       |         Male         |
|Observations    |----------------------|
|                |     HS    |  College  |
|----------------+-----------+-----------|
|Race            |           |           |
|----------------|           |           |
|White           |    150.00 |    715.00 |
|----------------+-----------+-----------|
|Black           |      9.00 |     53.00 |
---------------------------------------------
```

Making the Table Fit on the Page

It doesn't take very long in working with PROC TABULATE to encounter the dreaded "CONTINUED" message. This means that your table is too big to fit on the page. PROC TABULATE breaks up the table into page-sized pieces, prints them on multiple pages, and displays the "CONTINUED" message to let you know what happened.

Unfortunately, PROC TABULATE doesn't always pick the best places to split up a table. It's always better if you can get the entire table onto a single page. This chapter shows a number of tips and tricks for squeezing the maximum amount of information onto a page.

The Example

Income	Mgmt.		Prof.		Tech.		Sales	
	N	Mean	N	Mean	N	Mean	N	Mean
White								
Female	999	$99,999	999	$99,999	999	$99,999	999	$99,999
Male	999	$99,999	999	$99,999	999	$99,999	999	$99,999
Total	999	$99,999	999	$99,999	999	$99,999	999	$99,999
Black								
Female	999	$99,999	999	$99,999	999	$99,999	999	$99,999
Male	999	$99,999	999	$99,999	999	$99,999	999	$99,999
Total	999	$99,999	999	$99,999	999	$99,999	999	$99,999
Other								
Female	999	$99,999	999	$99,999	999	$99,999	999	$99,999
Male	999	$99,999	999	$99,999	999	$99,999	999	$99,999
Total	999	$99,999	999	$99,999	999	$99,999	999	$99,999

Reformatting CLASS Variables

Tables become too big for the page for two reasons–either there are too many columns to fit on the page, or there are too many rows to fit on the page. One way to fit more information on a single page is to get rid of excess rows or columns.

For example, the following code produces a table that is too wide to fit on a single page:

```
PROC TABULATE DATA=TEMP;
   CLASS RACE FULLTIME;
   VAR NUMKIDS;
   TABLE FULLTIME, RACE=' '*NUMKIDS=' '*(N MEAN)
      / BOX='Number of Children';
RUN;
```

This table is much too wide to fit on one page. If you look at Output 13.1, you can see that there is a "(CONTINUED)" label following the first page. The reason this table is so wide is that there are five values for RACE. The quickest way to shrink this table would be to cut down on the number of columns.

To see if this is possible, we need to look at the data. If there are categories with very few observations, these may be candidates to be combined. In this example, it turns out that most of the observations have a race of either 'White' or 'Black'. The other three categories ('Hispanic', 'Asian', and 'Native American') each have very few observations. These three categories could be combined to save space.

The way to combine the categories is to give the variable a new format. In the PROC FORMAT statement below, three of the race categories are assigned to 'Other'. Then the PROC TABULATE code can be run again, with the addition of a FORMAT statement to assign the new format to RACE.

```
PROC FORMAT;
   VALUE NEWRACE 1='White'
                 2='Black'
                 3-5='Other';
RUN;
PROC TABULATE DATA=TEMP;
   CLASS RACE FULLTIME;
   FORMAT RACE NEWRACE.;
   VAR NUMKIDS;
   TABLE FULLTIME, RACE=' '*NUMKIDS=' '*(N MEAN)
      / BOX='Number of Children';
RUN;
```

In Output 13.2, you can see that the table is now much smaller. It still doesn't fit on a single page, but we're getting closer. Now instead of two full pages of output, we have a page and one extra column.

With a little more work, we can get the whole table onto a single page, as you will see in the next example.

Output 13.1

Number of Children	White		Black		Hispanic
	N	MEAN	N	MEAN	N
Fulltime					
Full time	1527.00	1.52	139.00	1.60	67.00
Part time	173.00	1.54	8.00	1.50	12.00

(CONTINUED)

Number of Children	Hispanic	Asian		Native Amer.	
	MEAN	N	MEAN	N	MEAN
Fulltime					
Full time	1.58	70.00	1.49	44.00	1.11
Part time	1.33	8.00	1.00	8.00	1.25

Output 13.2

Number of Children	White		Black		Other
	N	MEAN	N	MEAN	N
Fulltime					
Full time	1527.00	1.52	139.00	1.60	181.00
Part time	173.00	1.54	8.00	1.50	28.00

(CONTINUED)

Number of Children	Other
	MEAN
Fulltime	
Full time	1.43
Part time	1.21

Reducing the Space Used by Row Headings

If your table still doesn't fit on the page after limiting the number of rows and columns as much as possible, the next trick to try is to reduce the space used by the row headings. Remember from Chapter 10 that you can adjust the space used by the row headings by using the RTS= option in the TABLE statement.

For this example, we will continue with our table from the previous example. If you look back at Output 13.2, you will notice that there's a lot of white space in the row headings. They are much wider than they need to be. We can adjust this by making the row headings just as wide as they have to be, but no more.

To figure out the appropriate width, we need to look for the longest row heading. This turns out to be the variable label 'Employment'. This label is 10 spaces wide. So the following code resets the row heading to 10 spaces wide by using RTS=10:

```
PROC TABULATE DATA=TEMP;
   CLASS RACE FULLTIME;
   FORMAT RACE NEWRACE.;
   VAR NUMKIDS;
   TABLE FULLTIME, RACE=' '*NUMKIDS=' '*(N MEAN)
      / BOX='Number of Children' RTS=10;
RUN;
```

But if you look at Output 13.3, you can see that something went wrong. The 10 spaces we allowed for the row heading were not enough.

This is because PROC TABULATE uses two of the spaces in this row heading for the dividers (|) that make up the left and right borders of the row heading. In order to calculate RTS settings, you need to take these special characters into account.

So for our table, we take the 10 spaces we need for the widest heading and add one space for each of the divider characters to get a RTS setting of 12 spaces.

```
PROC TABULATE DATA=TEMP;
   CLASS RACE FULLTIME;
   FORMAT RACE NEWRACE.;
   VAR NUMKIDS;
   TABLE FULLTIME, RACE=' '*NUMKIDS=' '*(N MEAN)
      / BOX='Number of Children' RTS=12;
RUN;
```

The resulting table is shown in Output 13.4. Now the row headings have just enough space to hold the widest label. In this example, we had two divider characters, so we added two spaces to the RTS setting. If we had nested row variables, there would be more dividers, and we'd have to add more spaces to the RTS setting to allow for them.

Our final table now fits on a single page. But if it didn't, there are still other tricks you can use to make a table narrower, as you will see in the examples to follow.

Output 13.3

```
--------------------------------------------------------------------------------
|Number  |         White            |         Black            |         Other            |
|of      |--------------------------+--------------------------+------------------------- |
|Children|    N     |    MEAN       |    N     |    MEAN       |    N     |    MEAN       |
|--------+----------+---------------+----------+---------------+----------+---------------|
|Employm-|          |               |          |               |          |               |
|ent     |          |               |          |               |          |               |
|--------|          |               |          |               |          |               |
|Full    |          |               |          |               |          |               |
|time    |   1527.00|           1.52|    139.00|           1.60|    181.00|           1.43|
|--------+----------+---------------+----------+---------------+----------+---------------|
|Part    |          |               |          |               |          |               |
|time    |    173.00|           1.54|      8.00|           1.50|     28.00|           1.21|
--------------------------------------------------------------------------------
```

Output 13.4

```
----------------------------------------------------------------------------------
|Number of |        White           |        Black           |        Other           |
|Children  |------------------------+------------------------+------------------------|
|          |    N    |    MEAN      |    N    |    MEAN      |    N    |    MEAN      |
|----------+---------+--------------+---------+--------------+---------+--------------|
|Employment|         |              |         |              |         |              |
|----------|         |              |         |              |         |              |
|Full time |  1527.00|          1.52|   139.00|          1.60|   181.00|          1.43|
|----------+---------+--------------+---------+--------------+---------+--------------|
|Part time |   173.00|          1.54|     8.00|          1.50|    28.00|          1.21|
----------------------------------------------------------------------------------
```

Reducing the Column Widths

The next step to take in reducing the size of your table is to cut down on all of that wasted space in the columns. Columns only need to be wide enough to hold the largest value that will appear in each column. The following code takes our example from the previous page, and re-creates it using much smaller column widths.

```
PROC TABULATE DATA=TEMP;
   CLASS RACE FULLTIME;
   FORMAT RACE NEWRACE.;
   VAR NUMKIDS;
   TABLE FULLTIME, RACE=' '*NUMKIDS=' '*(N*F=6. MEAN*F=5.1)
      / BOX='Number of Children' RTS=12;
RUN;
```

In Output 13.5, you can see that our table from the previous example now fits easily on the page. In fact, if we wanted to have narrower columns, we could have made it even smaller.

But you need to be careful with very narrow column widths. For example, the following table has two statistics, N and MEAN. We know from prior runs that N is never larger than four digits, and MEAN is never larger than 5 digits, so we can give them the narrowest possible formats 4 and 5 spaces wide.

```
PROC TABULATE DATA=TEMP;
   CLASS RACE FULLTIME;
   FORMAT RACE NEWRACE.;
   VAR INCOME;
   TABLE FULLTIME, RACE=' '*INCOME=' '*(N*F=4. MEAN*F=DOLLAR5.)
      / BOX='Income' RTS=12;
RUN;
```

If you look at Output 13.6, you can see that these formats are just wide enough to display the numbers. But there are two problems. First, with no space before the number in each column, it's hard to see where the columns are and hard to read the numbers. Second, the format for the means was too narrow to leave room for the dollar sign or comma associated with the DOLLAR5. format, so they were dropped.

PROC TABULATE will always try to make your values fit in the width you have specified, even if the column is so narrow that the values have to be shown in scientific notation. But a better way to build your tables is to allow a little bit of extra space in each column, as in the following code:

```
PROC TABULATE DATA=TEMP;
   CLASS RACE FULLTIME;
   FORMAT RACE NEWRACE.;
   VAR INCOME;
   TABLE FULLTIME, RACE=' '*INCOME=' '*(N*F=6. MEAN*F=DOLLAR8.)
      / BOX='Income' RTS=12;
RUN;
```

In Output 13.7, the table now has just enough room to hold the full format for each variable, and there's at least one space in each column to break up the text and make the table readable.

Output 13.5

```
------------------------------------------------------
|Number of |   White    |   Black   |   Other    |
|Children  |-----------+-----------+----------- |
|          |  N  |MEAN |  N  |MEAN |  N  |MEAN |
|----------+-----+-----+-----+-----+-----+-----|
|Employment|     |     |     |     |     |     |
|----------|     |     |     |     |     |     |
|Full time |  1527| 1.5|  139|  1.6|  181| 1.4|
|----------+-----+-----+-----+-----+-----+-----|
|Part time |  173| 1.5|    8|  1.5|   28| 1.2|
------------------------------------------------------
```

Output 13.6

```
---------------------------------------------------
|Income   |   White   |   Black   |   Other    |
|         |-----------+-----------+-----------|
|         | N  |MEAN | N  |MEAN | N  |MEAN |
|---------+----+-----+----+-----+----+-----|
|Employment|    |     |    |     |    |     |
|----------|    |     |    |     |    |     |
|Full time |1527|39864| 139|47531| 181|45216|
|----------+----+-----+----+-----+----+-----|
|Part time | 173|38148|   8|34340|  28|31030|
---------------------------------------------------
```

Output 13.7

```
------------------------------------------------------------
|Income    |    White    |    Black     |    Other     |
|          |-------------+--------------+--------------|
|          | N  | MEAN  | N  |  MEAN  | N  |  MEAN  |
|----------+----+-------+----+--------+----+--------|
|Employment|    |       |    |        |    |        |
|----------|    |       |    |        |    |        |
|Full time | 1527| $39,864|  139| $47,531|  181| $45,216|
|----------+----+-------+----+--------+----+--------|
|Part time |  173| $38,148|    8| $34,340|   28| $31,030|
------------------------------------------------------------
```

Reformatting the Column Variables

In the previous example, we reduced column widths to get a table to fit on the page. We made each column as narrow as possible given the size of the values we needed to display in the table. For example, the following table uses a very narrow FORMAT of 6.1 to squeeze a lot of columns on the page:

```
PROC TABULATE DATA=TEMP FORMAT=6.1;
    CLASS RACE OCCUP GENDER;
    FORMAT RACE NEWRACE.;
    VAR NUMKIDS;
    TABLE RACE, GENDER=' '*OCCUP=' '*NUMKIDS=' '*(MEAN)
        / BOX='Number of Children' RTS=12;
RUN;
```

The problem with this setting is apparent if you look at Output 13.8. The columns are wide enough to hold the table *values*, but they are not nearly wide enough to hold the column *headings*. When you select a format, it has to be wide enough not just for the table values but also for the column heading labels. If the format is not wide enough, PROC TABULATE splits up the labels and puts in hyphens as best it can, but the result is not very pretty. We can improve upon this by reformatting the class variable to create shorter labels.

There are two ways we can reformat the value labels. One is to keep the existing labels, but put in hyphens to tell PROC TABULATE where to put in the line breaks. The second way is to shorten the labels so that they fit in the six spaces allowed for each column heading.

```
PROC FORMAT;
    VALUE OCC1FT  1='Manag-ement'
                  2='Profes-sional'
                  3='Tech-nical'
                  4='Sales';
    VALUE OCC2FT  1='Mgmt.'
                  2='Prof.'
                  3='Tech.'
                  4='Sales';
    RUN;
```

The table shown in Output 13.9 uses the first format, OCC1FT. Because we put hyphens in the value labels, PROC TABULATE used these to break up the labels at the places we chose, instead of putting in a hyphen whenever it ran out of space.

 You need to be careful when using hyphenated labels with PROC TABULATE. If there is enough space in the heading to fit the whole label, then PROC TABULATE prints the entire label. Including the hyphen!

The table shown in Output 13.10 uses the second format, OCC2FT. These labels are a little harder to read, but they fit in the space allotted without any hyphens. If you can create meaningful abbreviations, this is the best way to reduce the size of your column headings (or row headings). If you can't create meaningful abbreviations, then you will be forced to use hyphens instead.

Output 13.8

Number of Children	Female				Male			
	Manag-erial	Profe-ssion-al	Techn-ical	Sales	Manag-erial	Profe-ssion-al	Techn-ical	Sales
	MEAN	MEAN	MEAN	MEAN	MEAN	MEAN	MEAN	MEAN
Race								
White	1.5	1.4	1.5	1.6	1.6	1.6	1.5	1.4
Black	1.5	1.6	2.3	1.5	1.9	1.3	0.5	1.5
Other	1.5	1.7	1.3	1.4	1.2	1.2	1.1	1.7

Output 13.9

Number of Children	Female				Male			
	Manag-ement	Profes-sional	Tech-nical	Sales	Manag-ement	Profes-sional	Tech-nical	Sales
	MEAN	MEAN	MEAN	MEAN	MEAN	MEAN	MEAN	MEAN
Race								
White	1.5	1.4	1.5	1.6	1.6	1.6	1.5	1.4
Black	1.5	1.6	2.3	1.5	1.9	1.3	0.5	1.5
Other	1.5	1.7	1.3	1.4	1.2	1.2	1.1	1.7

Output 13.10

Number of Children	Female				Male			
	Mgmt.	Prof.	Tech.	Sales	Mgmt.	Prof.	Tech.	Sales
	MEAN	MEAN	MEAN	MEAN	MEAN	MEAN	MEAN	MEAN
Race								
White	1.5	1.4	1.5	1.6	1.6	1.6	1.5	1.4
Black	1.5	1.6	2.3	1.5	1.9	1.3	0.5	1.5
Other	1.5	1.7	1.3	1.4	1.2	1.2	1.1	1.7

Moving the Statistics

So what do you do when you've made the row headings as narrow as possible, and squeezed all of the extra space out of your column widths, and the table is still too wide for the page? The next step is to look for things in the column dimension that could be moved to the row dimension.

For example, take a look at the following table. Notice how the F= and RTS= options have been used to minimize wasted space in the row headings and columns.

```
PROC TABULATE DATA=TEMP;
   CLASS OCCUP FULLTIME;
   VAR INCOME;
   TABLE FULLTIME, INCOME*OCCUP=' '*
      (N*F=5. MEAN*F=DOLLAR8. STD*F=DOLLAR8.)
      / RTS=12;
RUN;
```

But in Output 13.11, you can see that despite our best efforts, this table does not fit on a single page. There are just too many columns. One way to reduce the number of columns is to move the statistics from the column dimension to the row dimension. This way the statistics would repeat down the page instead of across the page. We'd end up with only four columns and a lot more rows. As long as the rows will all fit on the page, this is a great solution.

The following code produces the revised table. The only change that has been made is to move the section in parentheses that holds the statistics from the column dimension to the row dimension.

```
PROC TABULATE DATA=TEMP;
   CLASS OCCUP FULLTIME;
   VAR INCOME;
   TABLE FULLTIME*(N*F=5. MEAN*F=DOLLAR8. STD*F=DOLLAR8.),
      INCOME*OCCUP=' ' / RTS=12;
RUN;
```

The resulting table is shown in Output 13.12. As you can see, we have a problem. The narrow RTS= and F= settings that worked just fine when the statistics were in the columns don't work so well with statistics in rows. There is not enough space for the row headings or the column headings. However, the overall table is much narrower now, so we've got room to work with. We can double the RTS setting now that there are two layers of row heading, and we can expand the column formats to a width of 12, which will hold the widest of the column headings.

```
PROC TABULATE DATA=TEMP;
   CLASS OCCUP FULLTIME;
   VAR INCOME;
   TABLE FULLTIME*(N*F=12. MEAN*F=DOLLAR12. STD*F=DOLLAR12.),
      INCOME*OCCUP=' ' / RTS=24;
RUN;
```

Now our table, shown in Output 13.13, fits easily on the page.

Output 13.11

		Income						
		Managerial			Professional			Technical	
	N	MEAN	STD	N	MEAN	STD	N	MEAN
Employment								
Full time	670	$47,301	$33,176	745	$40,808	$25,798	118	$31,029
Part time	74	$37,303	$26,221	89	$41,894	$24,399	16	$31,723

(CONTINUED)

Output 13.12

		Income			
		Manager-ial	Profess-ional	Technic-al	Sales
Emp-loy-ment					
Full time	N	670	745	118	373
	MEAN	$47,301	$40,808	$31,029	$33,023
	STD	$33,176	$25,798	$17,032	$27,384
Part time	N	74	89	16	39
	MEAN	$37,303	$41,894	$31,723	$27,998
	STD	$26,221	$24,399	$22,091	$28,593

Output 13.13

		Income			
		Managerial	Professional	Technical	Sales
Employment					
Full time	N	670	745	118	373
	MEAN	$47,301	$40,808	$31,029	$33,023
	STD	$33,176	$25,798	$17,032	$27,384
Part time	N	74	89	16	39
	MEAN	$37,303	$41,894	$31,723	$27,998
	STD	$26,221	$24,399	$22,091	$28,593

Rotating the Table

Okay, so you've tried absolutely everything, and the table is still too wide for the page. What next? Well, one option is to turn the page. If you've got a short and wide table, why try to print it on a tall and narrow piece of paper? Instead, you can change the page orientation from portrait to landscape. Then your table will fit.

To change the page orientation, you need to do two things: First, you need to reset your line and page size options. Second, you need to select landscape orientation when you print your output. How you do this will depend on your operating system.

To set up PROC TABULATE for a landscape-oriented table, you need to widen the LINESIZE and shorten the PAGESIZE (make your page wider and shorter). The actual settings to use will vary depending on your margin settings and printer, but try the option settings in the following code as a starting point:

```
OPTIONS LS=132 PS=50;
PROC TABULATE DATA=TEMP;
   CLASS RACE OCCUP GENDER;
   FORMAT RACE NEWRACE.;
   VAR NUMKIDS;
   TABLE RACE, GENDER=' '*OCCUP=' '*NUMKIDS=' '*MEAN=' '
      / BOX='Number of Children' RTS=12;
RUN;
```

The resulting table is shown in Output 13.14. The output has been reduced in size to get it to fit on the page, but you can see that if this table were printed in landscape orientation, it would easily fit on a page.

This solution works great if you don't mind producing reports that have tables facing sideways. However, if you like all of your tables to be printed in portrait orientation, you do have another option. Instead of rotating the table, why not reverse the rows and columns? If you make your rows into columns and your columns into rows, you will end up with a table that is tall instead of wide. It should then fit on the page.

The following code takes our example table and reverses the rows and columns. Basically, the column dimension and the row dimension were swapped. Two other changes were made: the RTS setting was widened to make room for the nested headings, and the ROW=FLOAT option was turned on to remove the empty row headings.

```
PROC TABULATE DATA=TEMP;
   CLASS RACE OCCUP GENDER;
   FORMAT RACE NEWRACE.;
   VAR NUMKIDS;
   TABLE GENDER=' '*OCCUP=' '*NUMKIDS=' '*MEAN=' ', RACE
      / BOX='Number of Children' RTS=26 ROW=FLOAT;
RUN;
```

As you can see in Output 13.15, the table now fits easily on a portrait-oriented page. This may not work with all tables. Sometimes your column headings don't work as row headings, but it's always worth a try if you're having trouble getting the table to fit on a page.

Output 13.14

Number of Children	Female				Male			
	Managerial	Professional	Technical	Sales	Managerial	Professional	Technical	Sales
Race								
White	1.52	1.40	1.51	1.59	1.60	1.60	1.52	1.36
Black	1.53	1.56	2.33	1.47	1.90	1.34	0.50	1.54
Other	1.47	1.70	1.29	1.42	1.23	1.16	1.14	1.68

Output 13.15

Number of Children		Race		
		White	Black	Other
Female	Managerial	1.52	1.53	1.47
	Professional	1.40	1.56	1.70
	Technical	1.51	2.33	1.29
	Sales	1.59	1.47	1.42
Male	Managerial	1.60	1.90	1.23
	Professional	1.60	1.34	1.16
	Technical	1.52	0.50	1.14
	Sales	1.36	1.54	1.68

Removing a Heading by Putting It in the Table BOX

The previous examples all showed tables that were too wide for the page, but what about tables that are too long to fit on the page?

The following example shows a table that has too many rows to fit on a single page:

```
PROC TABULATE DATA=TEMP;
   CLASS YEAR GENDER OCCUP;
   FORMAT RACE NEWRACE.;
   VAR INCOME;
   TABLE YEAR*(GENDER ALL),
      OCCUP=' '*INCOME*(N*F=5. MEAN*F=DOLLAR8.);
   RUN;
```

In Output 13.16, you can see that this table is too long to fit on the page. If you look at the part of the table that spilled over onto the second page, it appears that five rows of the table did not fit on the first page. Actually, there was only has one row that wouldn't fit. The four rows taken up by the row and column headings had to be carried over onto the second page so that the one remaining row of actual data would make sense.

So this table is very close to fitting on a single page. All we have to do to get this table to fit is to reduce the number of rows by one.

We could do this by dropping one year of data from the table, but why lose data if we don't have to. A better approach is to get rid of one row in the column headings. In looking at the table, we can see that the value labels for each occupation can't be removed because we wouldn't know what each column represented. Similarly, we can't get rid of the statistic labels because we would not be able to tell the two statistics apart. The best candidate for removal is the label 'Income'. We can take the label for INCOME and put it in the table BOX.

```
PROC TABULATE DATA=TEMP;
   CLASS YEAR GENDER OCCUP;
   FORMAT RACE NEWRACE.;
   VAR INCOME;
   TABLE YEAR*(GENDER ALL),
      OCCUP=' '*INCOME=' '*(N*F=5. MEAN*F=DOLLAR8.)
      / BOX='Income';
   RUN;
```

The resulting table is not shown, but it does fit on a single page. And the 'Income' label in the table BOX saves space but still makes it clear what this table is all about.

Any time you have a table that is too long for the page, the first place to look for space to save is in the column headings. Frequently you have excess labels that can be removed. If your columns are 'Male' and 'Female', then a label of 'Gender' is redundant and can be removed. Or if your columns all use the same analysis variable, such as 'Income' above, the label can be moved to the BOX. Finally, if your columns all use the same statistic, then the statistic can be moved to the BOX as well. If our example table only used the MEAN statistic, then we could have labeled the BOX 'Mean Income' and gotten rid of two layers of column headings.

Output 13.16

| | | Managerial | | Professional | | Technical | | Sales | |
| | | Income | | Income | | Income | | Income | |
		N	MEAN	N	MEAN	N	MEAN	N	MEAN
Year	Gender								
1994	Gender								
	Female	7	$100,475	25	$78,468	.	.	1	$59,125
	Male	.	.	15	$61,566	1	$60,058	1	$10,035
	ALL	7	$100,475	40	$72,129	1	$60,058	2	$34,580
1995	Gender								
	Female	19	$78,116	66	$69,307	1	$91,529	5	$32,343
	Male	6	$41,428	28	$39,896	1	$19,615	4	$4,751
	ALL	25	$69,311	94	$60,546	2	$55,572	9	$20,080
1996	Gender								
	Female	211	$56,143	165	$44,430	33	$31,161	103	$51,075
	Male	150	$32,004	251	$30,702	40	$25,978	72	$24,942
	ALL	361	$46,113	416	$36,147	73	$28,321	175	$40,323
1997	Gender								
	Female	60	$43,027	24	$38,568	19	$35,152	44	$35,247
	Male	65	$31,465	27	$23,664	17	$23,585	31	$18,372
	ALL	125	$37,015	51	$30,678	36	$29,690	75	$28,272
1998	Gender								
	Female	70	$41,298	13	$35,670	15	$36,887	75	$34,964
	Male	63	$24,239	22	$25,136	13	$24,048	88	$13,354

(CONTINUED)

| | | Managerial | | Professional | | Technical | | Sales | |
| | | Income | | Income | | Income | | Income | |
		N	MEAN	N	MEAN	N	MEAN	N	MEAN
Year	ALL								
1998			133	$33,217	35	$29,048	28	$30,926	163	$23,297

Removing the Row Separators

Another way to make a tall table shorter is to remove the row separators. This has the potential to save a lot of space, but it also has the potential to make your table very hard to read.

In an ordinary table with row separators, each row is actually two lines of text. If you use the NOSEPS option and remove the row separators, you can create rows that use only one line of text. This means that the height of the body of your table will be cut in half.

For example, the table produced by the following code would be impossibly tall with row separators. It would be unlikely to fit on a single page. However, using the NOSEPS option, the table will be much shorter.

```
PROC TABULATE DATA=TEMP NOSEPS;
   CLASS YEAR AGE;
   VAR SAT;
   TABLES YEAR=' ',
      SAT=' '*AGE='Age Group'*MEAN=' '*F=6.2
      / RTS=15 BOX='Mean Satisfaction Score';
RUN;
```

If you look at Output 13.17, you can see that with the NOSEPS option turned on, the table easily fits on the page. However, there are so many numbers in the table it is hard to read. If you're going to use NOSEPS, you need to find ways to break up your tables.

One way to do this is to put spaces in the table every 10 years. To do this, we create a new CLASS variable called DECADE.

```
DATA TEMP2;
   SET TEMP;
   DECNUM=FLOOR(YEAR/10)*10;
   DECADE=DECNUM||'s';
RUN;
PROC TABULATE DATA=TEMP2 NOSEPS;
   CLASS YEAR AGE DECADE;
   VAR SAT;
   TABLES DECADE=' '*YEAR=' ',
      SAT=' '*AGE='Age Group'*MEAN=' '*F=6.2
      / RTS=15 INDENT=3 BOX='Mean Satisfaction Score';
RUN;
```

The revised table, shown in Output 13.18, is a little easier to read. With breaks every 10 rows, it's easier for the eye to scan across the rows and columns. Also, notice how the INDENT option is used to make the row headings easier to read. When you remove the row separators, you should always use INDENT if you have nested CLASS variables. Otherwise, the row headings are just too hard to read.

If we had not used INDENT in this example, PROC TABULATE would have listed the DECADE variable side by side with the first year of each decade. Then we wouldn't have ended up with any spaces in the table between each decade. To get the extra white space, you need to use INDENT.

Output 13.17

```
-----------------------------------
|Mean          |     Age Group       |
|Satisfaction  |------------------- |
|Score         |20-39 |40-59 | 60+   |
|-------------+------+------+------ |
|1975          |  1.33|  1.23|  1.19 |
|1976          |  1.39|  1.43|  1.32 |
|1977          |  1.44|  1.81|  1.70 |
|1978          |  1.33|  1.27|  1.23 |
|1979          |  1.79|  1.55|  1.48 |
|1980          |  1.74|  1.77|  1.25 |
|1981          |  1.45|  1.09|  1.02 |
|1982          |  1.31|  1.61|  1.30 |
|1983          |  1.89|  1.60|  1.17 |
|1984          |  1.57|  1.47|  0.90 |
|1985          |  1.34|  1.00|  1.16 |
|1986          |  1.42|  0.83|  1.55 |
|1987          |  1.43|  1.21|  1.51 |
|1988          |  1.41|  1.31|  1.56 |
|1989          |  1.67|  1.50|  1.40 |
|1990          |  1.41|  1.60|  1.92 |
|1991          |  1.31|  1.59|  1.50 |
|1992          |  1.56|  1.46|  1.25 |
|1993          |  1.42|  2.13|  0.88 |
|1994          |  1.35|  1.40|  1.40 |
|1995          |  1.39|  1.49|  1.13 |
|1996          |  1.33|  1.45|  1.48 |
|1997          |  1.49|  1.70|  1.18 |
|1998          |  1.55|  1.57|  1.54 |
-----------------------------------
```

Output 13.18

```
-----------------------------------
|Mean          |     Age Group       |
|Satisfaction  |------------------- |
|Score         |20-39 |40-59 | 60+   |
|-------------+------+------+------ |
|1970's        |      |      |       |
|   1975       |  1.33|  1.23|  1.19 |
|   1976       |  1.39|  1.43|  1.32 |
|   1977       |  1.44|  1.81|  1.70 |
|   1978       |  1.33|  1.27|  1.23 |
|   1979       |  1.79|  1.55|  1.48 |
|1980's        |      |      |       |
|   1980       |  1.74|  1.77|  1.25 |
|   1981       |  1.45|  1.09|  1.02 |
|   1982       |  1.31|  1.61|  1.30 |
|   1983       |  1.89|  1.60|  1.17 |
|   1984       |  1.57|  1.47|  0.90 |
|   1985       |  1.34|  1.00|  1.16 |
|   1986       |  1.42|  0.83|  1.55 |
|   1987       |  1.43|  1.21|  1.51 |
|   1988       |  1.41|  1.31|  1.56 |
|   1989       |  1.67|  1.50|  1.40 |
|1990's        |      |      |       |
|   1991       |  1.31|  1.59|  1.50 |
... remainder of table not shown ...
```

Changing the Font Size

In the previous examples, we've made wide tables narrower and tall tables shorter. We've rotated pages and tables, too. But sometimes, no matter what you do, the table is just too big for the page. There is one more option to take a look at. You can reduce the font size used in your table.

Output 13.19 shows a table that is too big in both directions. The edges of the table are cut off in this version of the output because there was no room to display them. It turns out that this table is just a little too big in both directions. If we could just squeeze one more column and a few more rows onto the page, this table would work as is.

Because the output looks pretty readable otherwise, we can fix this problem by reducing the font size. There are a number of ways you can do this. One way would be to cut and paste your output into a word processor, and then reduce the font size. This is a hassle because you now have to use two software packages to produce your output.

Another way would be to change the standard SAS font to a smaller point size (how you do this will vary by operating system). The problem with this is that this changes the font size for all output. A better way to change the font size is to use SAS/GRAPH to modify your output after it is produced.

You can do this without modifying your existing code. All you do is take your PROC TABULATE code and place it between two PROC PRINTTO calls. The first PRINTTO redirects the output to a designated file. The second PRINTTO is like an off switch. It closes this file and redirects your output back to wherever it usually goes. The following code sets this up for our example table:

```
FILENAME LISFILE 'C:\TEMP\TABOUT.LIS';
PROC PRINTTO PRINT=LISFILE NEW; RUN;

  PROC TABULATE DATA=TEMP;
    CLASS OCCUP RACE GENDER;
    VAR INCOME;
    TABLE RACE=' '*(GENDER=' ' ALL),
      OCCUP=' '*INCOME=' '*(N*F=8. MEAN*F=DOLLAR10.)
      / BOX='Income' ROW=FLOAT;
  RUN;

  PROC PRINTTO; RUN;
```

All of this is just setting things up; it's the next part of the code that is interesting. We're going to use the graphics procedure GPRINT to print the output file created by the PROC PRINTTO. To reduce the font size, you can use the GOPTION setting HTEXT. The default HTEXT is 1, so pick a smaller value to reduce the font.

```
GOPTIONS HTEXT=.8 DEVICE=devicename;
PROC GPRINT FILEREF=LISFILE; RUN;
```

As you can see from Output 13.20, the table is now 80% of the size of the original table, and it now fits (barely) on the page. Be patient, it may take a bit of experimenting with the HTEXT setting to find the value that will work for your table. You want to use the largest value that will just fit on a page.

Output 13.19

Income		Managerial		Professional		Technic	
		N	Mean	N	Mean	N	
White	Female	351	$55,775	291	$49,877	51	
	Male	268	$32,274	387	$31,758	62	
	All	619	$45,600	678	$39,535	113	
Black	Female	36	$66,257	34	$51,608	9	
	Male	21	$29,787	29	$33,928	2	
	All	57	$52,820	63	$43,470	11	
Hispanic	Female	19	$48,990	20	$59,438	1	
	Male	8	$32,377	10	$34,188	2	
	All	27	$44,068	30	$51,021	3	
Asian	Female	15	$52,582	13	$44,645	3	
							...

Output 13.20

Income		Managerial		Professional		Technical		Sales	
		N	MEAN	N	MEAN	N	MEAN	N	MEAN
White	Female	351	$55,775	291	$49,877	51	$35,183	200	$41,881
	Male	268	$32,274	387	$31,758	62	$24,135	166	$18,483
	ALL	619	$45,600	678	$39,535	113	$29,121	366	$31,269
Black	Female	36	$66,257	34	$51,608	9	$24,126	15	$52,846
	Male	21	$29,787	29	$33,928	2	$48,770	13	$17,337
	ALL	57	$52,820	63	$43,470	11	$28,606	28	$36,359
Hispanic	Female	19	$48,990	20	$59,438	1	$43,030	13	$53,213
	Male	8	$32,377	10	$34,188	2	$30,065	9	$20,540
	ALL	27	$44,068	30	$51,021	3	$34,387	22	$39,846
Asian	Female	15	$52,582	13	$44,645	3	$41,677	10	$48,677
	Male	14	$36,890	17	$31,101	3	$25,042	6	$43,769
	ALL	29	$45,006	30	$36,970	6	$33,359	16	$46,836
Native Amer.	Female	15	$42,317	10	$59,039	3	$46,047	3	$49,514
	Male	9	$31,899	11	$32,656	2	$19,512	4	$16,338
	ALL	24	$38,410	21	$45,219	5	$35,433	7	$30,556

Exporting the Table to Spreadsheet

If you have a big table that you know will require a lot of fine-tuning to get it to fit on a page, you may be better off using another tool. SAS is good at producing tables, and it has many ways of reformatting these tables, but you may find that a spreadsheet gives you more control.

To do this, you use SAS for its strength: computing the tables. Spreadsheet programs are not good at turning data into tables. Then, you use the spreadsheet program for its strengths: reformatting rows, columns, values, and headings, and creating high-resolution output.

The process of exporting PROC TABULATE tables is covered in detail in Chapter 16, "Outputting the Results," but here's a quick example to whet your appetite.

First, you need to create the table. Just as in the previous example, we use PROC PRINTTO to re-route our data to a file. Other changes include the PAGESIZE and LINESIZE settings, the NOSEPS option, and the FORMCHAR setting. Also notice the generous space allowed in the RTS setting and the format for the table values.

```
FILENAME LISFILE 'C:\TEMP\TABOUT.LIS';
PROC PRINTTO PRINT=LISFILE NEW; RUN;

   OPTIONS PS=250 LS=250 NOCENTER;
   PROC TABULATE DATA=TEMP NOSEPS FORMCHAR=',            ';
      CLASS OCCUP RACE GENDER;
      VAR INCOME;
      TABLE RACE=' '*(GENDER=' ' ALL),
         OCCUP=' '*INCOME=' '*MEAN=' '*F=14.
         / BOX='Mean Income' ROW=FLOAT RTS=30;
   RUN;

   PROC PRINTTO; RUN;
```

The output file produced by this code looks pretty strange. Output 13.21 shows the first four columns of this table. It has lots of blank space between the columns, and there are commas all over the place.

Now look at the same table again after it has been imported into a spreadsheet. Output 13.22 shows what you can do with more advanced formatting tools. The font has been changed to Arial Narrow to fit all of the columns on the page.

This example shows how to force a wide table to fit on a narrow page, but you could also use the formatting tools available in a spreadsheet program to make a tall table shorter. For example, you could reduce the row height slightly to get a few more rows on the page.

Note also how the use of boldface, underlining, and a larger title make this table readable, even without any grid lines. We could also add color to the table or put a company logo at the top.

This example is a brief demonstration. For the details on how to export your tables to a spreadsheet or word processor, see Chapter 16.

Output 13.21

,Mean Income		,	Managerial	, Professional	, Technical	,	Sales	,
,White	Female	,	55775,	49877,	35183,		41881,	
,	Male	,	32274,	31758,	24135,		18483,	
,	ALL	,	45600,	39535,	29121,		31269,	
,Black	Female	,	66257,	51608,	24126,		52846,	
,	Male	,	29787,	33928,	48770,		17337,	
,	ALL	,	52820,	43470,	28606,		36359,	
,Hispanic	Female	,	48990,	59438,	43030,		53213,	
,	Male	,	32377,	34188,	30065,		20540,	
,	ALL	,	44068,	51021,	34387,		39846,	
,Asian	Female	,	52582,	44645,	41677,		48677,	
,	Male	,	36890,	31101,	25042,		43769,	
,	ALL	,	45006,	36970,	33359,		46836,	
,Native Amer.	Female	,	42317,	59039,	46047,		49514,	
,	Male	,	31899,	32656,	19512,		16338,	
,	ALL	,	3410,	45219,	35433,		30556,	

Output 13.22

Mean Incomes by Occupation, Race, and Gender					
		Managerial	**Professional**	**Technical**	**Sales**
White	**Female**	$55,775	$49,877	$35,183	$41,881
	Male	$32,274	$31,758	$24,135	$18,483
	ALL	$45,600	$39,535	$29,121	$31,269
Black	**Female**	$66,257	$51,608	$24,126	$52,846
	Male	$29,787	$33,928	$48,770	$17,337
	ALL	$52,820	$43,470	$28,606	$36,359
Hispanic	**Female**	$48,990	$59,438	$43,030	$53,213
	Male	$32,377	$34,188	$30,065	$20,540
	ALL	$44,068	$51,021	$34,387	$39,846
Asian	**Female**	$52,582	$44,645	$41,677	$48,677
	Male	$36,890	$31,101	$25,042	$43,769
	ALL	$45,006	$36,970	$33,359	$46,836
Native Amer.	**Female**	$42,317	$59,039	$46,047	$49,514
	Male	$31,899	$32,656	$19,512	$16,338
	ALL	$38,410	$45,219	$35,433	$30,556

Splitting Up the Table

Perhaps this should have been the first example in the chapter. If you can't get your table to fit on a single page, maybe your problem is that you've got too much information in a single table.

As you have seen in the preceding examples, there's a lot you can do to cram a large volume of information into a single page. But there are some trade-offs. The more you crowd your table, the harder it is to read. The best solution may be to give in gracefully and redesign your table as a series of tables. That's what the page dimension is for.

The following code should be familiar. This table is similar to the previous examples in this chapter. It uses a number of techniques to minimize the size of the table. A CLASS variable is reformatted, many labels have been removed, the RTS= option keeps the row headings short, and F= format options are used to minimize the column width.

```
PROC TABULATE DATA=TEMP;
   CLASS OCCUP RACE GENDER FULLTIME;
   FORMAT RACE NEWRACE.;
   VAR INCOME;
   TABLE RACE=' '*(GENDER=' ' ALL),
      FULLTIME=' '*OCCUP=' '*INCOME=' '*(N*F=4. MEAN*F=DOLLAR8.)
      / BOX='Income' RTS=21 ROW=FLOAT;
RUN;
```

But if you look at Output 13.23, you can see that the table still will not fit on the page. This table is too wide for a single page. However, if you look at the column headings, you can see that all of the columns for the first value of the CLASS variable FULLTIME would fit on a page. It's just the second value of FULLTIME that won't fit.

We could split this table in half, using FULLTIME as the page variable. This would give us a two-page table that breaks sensibly, instead of having the awkward break partway through the second value of FULLTIME. The only change we need to make in the code is to take the FULLTIME=' ' out of the column dimension and list it at the beginning of the TABLE statement as the page variable (followed by a comma).

```
PROC TABULATE DATA=TEMP;
   CLASS OCCUP RACE GENDER FULLTIME;
   FORMAT RACE NEWRACE.;
   VAR INCOME;
   TABLE FULLTIME=' ',
      RACE=' '*(GENDER=' ' ALL),
      OCCUP=' '*INCOME=' '*(N*F=4. MEAN*F=DOLLAR8.)
      / BOX='Income' RTS=21 ROW=FLOAT;
RUN;
```

Now you can see in Output 13.24 that our two-page table looks much better. All of the information for full-time workers is on the first page, and all of the information for part-time workers is on the second page. It would have been nice to have a table that would fit on a single page, but because that is not possible, at least this way we end up with two readable pages.

Output 13.23

Income		Full time								Part time
		Managerial		Professional		Technical		Sales		Man-age-rial
		N	MEAN	N	MEAN	N	MEAN	N	MEAN	N
White	Female	317	$56,741	254	$49,941	42	$34,628	168	$42,017	23
	Male	227	$32,871	332	$31,427	49	$24,919	138	$18,624	34
	ALL	544	$46,780	586	$39,452	91	$29,400	306	$31,467	57
Black	Female	32	$69,987	30	$54,375	8	$26,589	13	$58,553	2
	Male	20	$30,991	25	$35,552	2	$48,770	9	$17,324	.
	ALL	52	$54,989	55	$45,819	10	$31,025	22	$41,687	2
Other	Female	39	$52,656	40	$56,464	7	$43,743	24	$52,347	5
	Male	23	$34,884	28	$33,189	5	$28,544	15	$29,009	8
	ALL	62	$46,063	68	$46,880	12	$37,410	39	$43,371	13

(CONTINUED)

Output 13.24

Full time

Income		Managerial		Professional		Technical		Sales	
		N	MEAN	N	MEAN	N	MEAN	N	MEAN
White	Female	317	$56,741	254	$49,941	42	$34,628	168	$42,017
	Male	227	$32,871	332	$31,427	49	$24,919	138	$18,624
	ALL	544	$46,780	586	$39,452	91	$29,400	306	$31,467
Black	Female	32	$69,987	30	$54,375	8	$26,589	13	$58,553
	Male	20	$30,991	25	$35,552	2	$48,770	9	$17,324
	ALL	52	$54,989	55	$45,819	10	$31,025	22	$41,687
Other	Female	39	$52,656	40	$56,464	7	$43,743	24	$52,347
	Male	23	$34,884	28	$33,189	5	$28,544	15	$29,009
	ALL	62	$46,063	68	$46,880	12	$37,410	39	$43,371

The second page of this output is not included here.

PART 4

Advanced Topics

Using Macros to Generate Tables

After putting in the time to design a complex table, you don't want to waste your time retyping all of that code to run the same table on different variables or to generate different statistics.

One of the most powerful tools in the SAS System is the ability to create macros to do repetitive tasks.

If you've got a program that produces a series of tables for the month of January, you can add a macro that lets you run the tables again for February (and March, April, and so on).

You can also write a macro that lets you vary the statistics shown in the table. You can produce a whole series of tables showing various statistics for the same group of variables.

Or if you're ambitious, you can write a generic PROC TABULATE macro that lets you specify the variables, statistics, and time period, and then produces a standard table format. This chapter shows you how to build macros for each of these tasks.

Please note that this chapter is not intended to teach you how to write SAS macros. The point of the chapter is to show you how to apply macro-writing skills to PROC TABULATE. If you have no experience with macros, you may find the examples somewhat difficult.

Repeating a Table for a Series of CLASS Variables

Have you ever produced a great table for your boss, only to hear: "That's just what I wanted, but could I see the same table for x, y, and z as well?" This is easy to do if you turn your table into a macro and run it over and over again with the new variables. This example shows how to run the same table again using different classification variables.

First, we'll build our basic table. This is a crosstabulation of data from an employer survey that shows the companies' mean number of employees broken down by full-time versus part-time status. The data is also categorized into rows by whether the companies are in the public or private sector. The original code was as follows:

```
PROC TABULATE DATA=TEMP;
   CLASS SECTOR FULLTIME GENDER;
   VAR NUMEMP;
   TABLE SECTOR, FULLTIME=' '*GENDER=' '
      *NUMEMP=' '*MEAN=' '*F=8.
      / BOX="Mean Number of Employees";
   TITLE "Original Table Showing Company Data";
RUN;
```

The output from this code is shown in Output 14.1. To turn this code into a macro, all you have to do is figure out what variable you want to be able to replace with a macro variable. In this case, we want to vary the row definition. We want to be able to replace SECTOR with some other variable.

The first thing to do is put the PROC TABULATE code inside a macro definition. Because this macro requires that a row variable be specified, we use a macro parameter called ROWVAR:

```
%MACRO TABLEIT(ROWVAR);
   PROC TABULATE DATA=TEMP;
      CLASS FULLTIME GENDER &ROWVAR;
      VAR NUMEMP;
      TABLE &ROWVAR, FULLTIME=' '*GENDER=' '
         *NUMEMP=' '*MEAN=' '*F=8.
         / BOX="Mean Number of Employees";
      TITLE "Macro-Generated Table Showing Company Data";
   RUN;
%MEND TABLEIT;
```

The only other changes made to the PROC TABULATE code were to replace each reference to the variable SECTOR with a reference to the macro parameter &ROWVAR. To test that the new code works, we can reproduce our original table by calling TABLEIT with SECTOR as the parameter. Then we can run the table again with a new row variable (UNION).

```
%TABLEIT(SECTOR);
%TABLEIT(UNION);
```

These tables are shown in Output 14.2.

Output 14.1

Original Table Showing Company Data

```
------------------------------------------------------------------
|Mean Number of Employees      |    Full time    |   Part time    |
|                              |-----------------+--------------- |
|                              |  Male | Female  |  Male | Female |
|------------------------------+-------+---------+-------+------- |
|Sector                        |       |         |       |        |
|----------------------------- |       |         |       |        |
|Private                       |   394 |    410  |  373  |    383 |
|------------------------------+-------+---------+-------+------- |
|Public                        |   732 |    648  |  907  |    873 |
------------------------------------------------------------------
```

Output 14.2

Macro-Generated Table Showing Company Data

```
------------------------------------------------------------------
|Mean Number of Employees      |    Full time    |   Part time    |
|                              |-----------------+--------------- |
|                              |  Male | Female  |  Male | Female |
|----------------------------- +-------+---------+-------+------- |
|Sector                        |       |         |       |        |
|----------------------------- |       |         |       |        |
|Private                       |   394 |    410  |  373  |    383 |
|------------------------------+-------+---------+-------+------- |
|Public                        |   732 |    648  |  907  |    873 |
------------------------------------------------------------------
```

Macro-Generated Table Showing Company Data

```
------------------------------------------------------------------
|Mean Number of Employees      |    Full time    |   Part time    |
|                              |-----------------+--------------- |
|                              |  Male | Female  |  Male | Female |
|----------------------------- +-------+---------+-------+------- |
|Union Status                  |       |         |       |        |
|----------------------------- |       |         |       |        |
|Non-Union                     |   514 |    450  |  505  |    437 |
|------------------------------+-------+---------+-------+------- |
|Union                         |   727 |    625  |  900  |    859 |
------------------------------------------------------------------
```

Repeating a Table for a Series of Statistics

Sometimes you need to produce a series of tables with the same variables and different statistics. A table that shows means is interesting, but you may also want to show a row percentage or maybe the mean plus the standard deviation. To get started, we use the same basic code from the previous example:

```
PROC TABULATE DATA=TEMP;
   CLASS SECTOR FULLTIME GENDER;
   VAR NUMEMP;
   TABLE SECTOR, FULLTIME=' '*GENDER=' '
      *NUMEMP=' '*MEAN=' '*F=8.
      / BOX="Mean Number of Employess";
   TITLE "Original Table Showing Company Data";
RUN;
```

This time, instead of replacing the row variable with a macro variable, we replace the statistic. This is a little more complicated because we will also have to change the BOX label and the format so that they will be appropriate for the new statistics. Yes, that's statistics plural. With a little work, we can create a macro that will replace the mean with a single statistic or with multiple statistics.

This macro needs to have three parameters. The first lists the new statistics. The second is the BOX label. The third is the format for the table values. The format is included as a macro variable because a format such as 8. that works for some statistics may not work for others.

```
%MACRO TABLEIT(THESTAT, THELABEL, THEFMT);
   PROC TABULATE DATA=TEMP;
        CLASS FULLTIME GENDER SECTOR;
        VAR NUMEMP;
        TABLE SECTOR, FULLTIME=' '*GENDER=' '
           *NUMEMP=' '*(&THESTAT)*F=&THEFMT
           / BOX="&THELABEL";
        TITLE "Macro-Generated Table Showing Company Data";
   RUN;
%MEND TABLEIT;
```

In this code, the reference to the statistic MEAN has been replaced with a reference to the macro parameter &THESTAT. The parameter is enclosed in parentheses so that it can contain more than one statistic keyword. Next, the format following the F= is replaced with the macro variable &THEFMT. Finally, the BOX label is changed to the macro parameter &THELABEL. This label must be enclosed in double quotation marks because it contains a macro variable.

Again, to test the new code, first we run it with parameters that should reproduce the original table from the previous example. Then we can run additional tables showing other statistics. These tables are shown in Output 14.3, 14.4, and 14.5.

```
%TABLEIT(MEAN,Mean Number of Employees,8.);
%TABLEIT(MEAN STD,Number of Employees (Mean and Std), 8.2);
%TABLEIT(PCTN<FULLTIME*GENDER>,Percent Employees by Sector, 8.1);
```

Output 14.3

```
            Macro-Generated Table Showing Company Data
            ---------------------------------------------------------
            |Mean Number of   |     Full time    |     Part time    |
            |Employees        |---------------+----------------|
            |                 | Male  | Female | Male  | Female |
            |                 |--------+--------+--------+--------|
            |                 | MEAN  | MEAN   | MEAN  | MEAN   |
            |-----------------+--------+--------+--------+--------|
            |Sector           |       |        |       |        |
            |-----------------|       |        |       |        |
            |Private          |   394 |   410  |   373 |   383  |
            |-----------------+--------+--------+--------+--------|
            |Public           |   732 |   648  |   907 |   873  |
            ---------------------------------------------------------
```

Output 14.4

```
            Macro-Generated Table Showing Company Data
--------------------------------------------------------------------------------
|Number of        |              Full time              |              Part time              |
|Employees (Mean  |-------------------------------------+-------------------------------------|
|and Std)         |       Male        |      Female     |       Male        |      Female     |
|                 |-----------------+-----------------+-----------------+-----------------|
|                 | MEAN  | STD     | MEAN  | STD     | MEAN  | STD     | MEAN  | STD     |
|-----------------+--------+--------+--------+--------+--------+--------+--------+--------|
|Sector           |       |        |        |        |        |        |        |        |
|-----------------|       |        |        |        |        |        |        |        |
|Private          | 394.44| 433.40 | 409.97| 423.96 | 372.88| 440.85 | 383.47| 402.90 |
|-----------------+--------+--------+--------+--------+--------+--------+--------+--------|
|Public           | 732.11| 382.39 | 647.99| 397.35 | 906.52| 269.53 | 872.76| 303.64 |
--------------------------------------------------------------------------------
```

Output 14.5

```
            Macro-Generated Table Showing Company Data
            ---------------------------------------------------------
            |Percent Employees |     Full time    |     Part time    |
            |by Sector         |---------------+----------------|
            |                  | Male  | Female | Male  | Female |
            |                  |--------+--------+--------+--------|
            |                  | PCTN  | PCTN   | PCTN  | PCTN   |
            |------------------+--------+--------+--------+--------|
            |Sector            |       |        |       |        |
            |------------------|       |        |       |        |
            |Private           |  47.4 |   41.6 |   5.7 |   5.3  |
            |------------------+--------+--------+--------+--------|
            |Public            |  38.1 |   47.7 |   5.1 |   9.1  |
            ---------------------------------------------------------
```

Repeating a Table for Various Time Periods - I

You can also use macros to repeat the same table for various subsets of the data. For example, you can run a table showing income figures for a particular month or year.

As in the previous examples, we start with a basic table. In this case, the table shows income by union status and full-time status. The original table is run on all observations in the data set:

```
PROC TABULATE DATA=TEMP;
    CLASS UNION FULLTIME;
    VAR INCOME;
    TABLE UNION=' ', FULLTIME=' '*INCOME=' '*MEAN=' ';
    TITLE "Mean Income (All Observations)";
RUN;
```

This output is shown in Output 14.6. To create a macro that to run this table for a particular time period, we need to add a WHERE clause to the procedure. But instead of just adding a WHERE clause for the desired time period, we can add a WHERE clause in which the time period is specified by a macro variable. Then, when you want to run the table for a different time period, all you have to do is change the macro call.

```
%MACRO TABLEIT(WHERECL,THELABEL);
    PROC TABULATE DATA=TEMP;
        WHERE &WHERECL;
        CLASS UNION FULLTIME;
        VAR INCOME;
        TABLE UNION=' ', FULLTIME=' '*INCOME=' '*MEAN=' ';
        TITLE "Mean Income (&THELABEL)";
    RUN;
%MEND TABLEIT;
```

Note that the macro also needs to have a parameter that gives the time period a label so that the table can have an appropriate title indicating the time frame selected in the WHERE clause. Without this label, the end user would have no idea what time frame the table represents.

Now we can call the macro and request a table for a particular month or year of data.

```
%TABLEIT('1JAN97'D<=DATE<='31JAN97'D,January 1997);
%TABLEIT(YEAR(DATE)=1997,1997);
```

The tables from these macros are shown in Output 14.7. This macro is so basic that it is not limited to selecting date ranges. Any valid WHERE clause that can be expressed as a macro variable can be used with this macro. For example, we can limit the table to a particular gender. The only trick to using macro variables in WHERE clauses is that there are limitations on using certain symbols (such as equal signs) in macro parameters. This can be avoided by changing the "=" to an "EQ" or by using the %STR macro function. The following code produces the table shown in Output 14.8:

```
%TABLEIT(GENDER EQ 2, Females);
```

Output 14.6

Mean Income (All Observations)

```
---------------------------------------------------------------
|                              | Full time | Part time |
|------------------------------+-----------+-----------|
|Non-Union                     |   33671.22|   32083.00|
|------------------------------+-----------+-----------|
|Union                         |   33967.03|   31485.55|
---------------------------------------------------------------
```

Output 14.7

Mean Income (January 1997)

```
---------------------------------------------------------------
|                              | Full time | Part time |
|------------------------------+-----------+-----------|
|Non-Union                     |   29640.87|   28687.56|
|------------------------------+-----------+-----------|
|Union                         |   32242.73|   28100.50|
---------------------------------------------------------------
```

Mean Income (1997)

```
---------------------------------------------------------------
|                              | Full time | Part time |
|------------------------------+-----------+-----------|
|Non-Union                     |   39229.90|   40086.19|
|------------------------------+-----------+-----------|
|Union                         |   36313.20|   34964.33|
---------------------------------------------------------------
```

Output 14.8

Mean Income (Females)

```
---------------------------------------------------------------
|                              | Full time | Part time |
|------------------------------+-----------+-----------|
|Non-Union                     |   25266.73|   24414.02|
|------------------------------+-----------+-----------|
|Union                         |   26817.98|   33740.62|
---------------------------------------------------------------
```

Repeating a Table for a Series of Time Periods - II

Sometimes, you want to build tables with so many different combinations of variables and time periods that even building a macro with parameters is too much work. It can take a long time to set up all the macro calls with the appropriate variable and time period selections. In this case, you may want to set up your macro so that it loops through the various combinations of variables and time periods.

For example, say you want to produce tables of income by gender for each of four classification variables, and you want to repeat these tables for each of 12 years. If you set up a query with parameters for the CLASS variable and year, you'd have to set up 48 (12x4) macro calls!

Alternatively, you can set up a macro like the one shown below:

```
%LET VAR1=Race;
%LET VAR2=Marital;
%LET VAR3=Occup;
%LET VAR4=Fulltime;
%MACRO TABLEIT;
   %DO Y=1985 %TO 1996;
      %DO V=1 %TO 4;
         PROC TABULATE DATA=TEMP;
            WHERE YEAR=&Y;
            CLASS &&VAR&V GENDER;
            VAR INCOME;
            TABLE &&VAR&V=' ',
               GENDER=' '*INCOME=' '*MEAN=' '*F=DOLLAR8. /
               ROW=FLOAT BOX="Mean Income";
            TITLE "Table of Mean Income by Gender and &&VAR&V for &Y";
         RUN;
      %END;
   %END;
%MEND TABLEIT;
```

In this macro, the four CLASS variables are set up as macro variables VAR1-VAR4, and then called by the macro one at a time as it cycles through the 12 years. The result is 48 tables from one macro call:

```
%TABLEIT;
```

The output is shown in Output 14.9. The output from the first iteration of the loops is a table for 1985 that shows INCOME by RACE and GENDER. The second iteration of the loops produces the same table, except by MARITAL and GENDER. The third loop (not shown) produces a table by OCCUP and GENDER. The fourth loop (not shown) produces a table by FULLTIME and GENDER.

For the fifth loop, the tables start over with the year set to 1986. This process continues until the 48th table (shown at the bottom of the facing page) is produced, showing mean INCOME for 1996 broken down by FULLTIME and GENDER.

Output 14.9

```
    Table of Mean Income by Gender and Race for 1985

    ---------------------------------------------------
    |Mean Income                 | Female |  Male  |
    |----------------------------+--------+------- |
    |White                       | $18,317| $30,078|
    |----------------------------+--------+------- |
    |Black                       | $20,896| $22,224|
    |----------------------------+--------+------- |
    |Other                       | $15,000|      . |
    ---------------------------------------------------
```

```
    Table of Mean Income by Gender and Marital for 1985

    ---------------------------------------------------
    |Mean Income                 | Female |  Male  |
    |----------------------------+--------+------- |
    |Married                     | $17,597| $32,438|
    |----------------------------+--------+------- |
    |Widowed                     | $17,407| $17,107|
    |----------------------------+--------+------- |
    |Divorced/Sep.               | $20,567| $30,552|
    |----------------------------+--------+------- |
    |Never Married               | $18,854| $20,197|
    ---------------------------------------------------
```

             ... (NEXT 45 TABLES NOT SHOWN) ...

```
    Table of Mean Income by Gender and Fulltime for 1996

    ---------------------------------------------------
    |Mean Income                 | Female |  Male  |
    |----------------------------+--------+------- |
    |Full time                   | $18,181| $30,819|
    |----------------------------+--------+------- |
    |Part time                   | $17,684| $28,239|
    ---------------------------------------------------
```

Creating a Generic Macro to Produce Tables

If you regularly produce tables with a similar design, you can create a generic macro to simplify the table production process. This allows you to set up a basic format for titles, footnotes, row variables, and column variables, and then use it in all of your tables.

To produce this generic macro, you start by creating the basic PROC TABULATE code. This is the part of the code that doesn't vary between your tables. Just leave placeholders for the parts that will change from table to table, which will be filled in later by macro variables.

```
PROC TABULATE DATA=TEMP;
    CLASS var1 var2;
    VAR var3;
    TABLE var1,
        var2 * var3 *(stat)*F=fmt
        / ROW=FLOAT BOX="text1";
    TITLE "text2";
RUN;
```

In this example, we have a basic table format with two classification variables: one for the rows and one for the columns. Then there is the analysis variable, a statistic, and a format. Finally, there are two text strings, one for the BOX label and one for the TITLE. The two text strings can be built from the other variables, so we end up with a macro that has five parameters: a row variable, column variable, analysis variable, format, and statistic.

By substituting macro variables for the placeholders in the code above, you end up with the macro below:

```
%MACRO TABLEIT(THEROW,THECOL,THEVAR,THEFMT,THESTAT);
    PROC TABULATE DATA=TEMP;
        CLASS &THEROW &THECOL;
        VAR &THEVAR;
        TABLE &THEROW=' ',
            &THECOL=' '*&THEVAR=' '*(&THESTAT)*F=&THEFMT
            / ROW=FLOAT BOX="&THEVAR";
        TITLE "Table of &THEVAR (&THESTAT) by &THEROW and &THECOL";
    RUN;
%MEND TABLEIT;
```

To call the macro, you put in your choice of variables, format, and statistic. Below are three macro calls that produce three different tables. The resulting tables are shown in Output 14.10.

```
%TABLEIT(GENDER,RACE,INCOME,8.,N);
%TABLEIT(MARITAL,EMPLOYED,AGE,8.1,MEAN STD);
%TABLEIT(OCCUP,FULLTIME,HOURLY,DOLLAR8.2,MAX);
```

Although the content of the tables is quite different, the structure remains the same. This can be useful when you have a number of programmers producing tables and want to encourage standardization. This example is fairly simple, but you can use this approach to build more complex table macros.

Output 14.10

Table of Income (N) by Gender and Race

Income	White	Black	Other
	N	N	N
Female	3353	89	8
Male	2908	76	7

Table of Age (Mean Std) by Marital and Employed

Age	Unemployed		Employed	
	MEAN	STD	MEAN	STD
Married	60.2	16.5	43.2	10.7
Widowed	76.1	9.3	57.8	10.3
Divorced/Sep.	54.9	15.6	43.5	10.1
Never Married	51.6	20.6	34.1	9.0

Table of Hourly (Max) by Occup and Fulltime

Hourly	Full time	Part time
	MAX	MAX
Managerial	$126.00	$76.75
Professional	$101.50	$69.69
Technical	$45.76	$50.25
Sales	$85.31	$55.40
Clerical	$54.53	$34.12
Services	$38.71	$65.01
Manufacturing	$68.51	$43.84
Farming	$38.67	$34.20

CHAPTER 15

PROC TABULATE Tricks: How to Cheat to Create Complex Tables

Most of the time, there is a straightforward way to produce the table you want using PROC TABULATE. But sometimes, you want to do something that just doesn't seem to be listed in the documentation.

This chapter introduces a number of programming tricks to get PROC TABULATE to do things it wasn't designed to do. Sometimes this involves creative formatting or labeling of your output. Other times you need to use DATA steps or other SAS procedures to manipulate your data before it is passed to PROC TABULATE.

The important thing to remember about most of the tricks introduced in this chapter is that they need to be documented in your code. If another programmer later uses your programs, they're going to find these techniques highly confusing unless you explain what's going on.

Of course this chapter cannot solve all of your PROC TABULATE challenges. There are always things that PROC TABULATE just can't do. You may need to use a different tool to create your desired output. Chapter 20, "Limitations of PROC TABULATE and How to Get Around Them," suggests other SAS procedures that are better suited to certain tasks. Also, Chapter 16, "Outputting the Results," shows you how to export your PROC TABULATE output so that you can make further modifications using other software (spreadsheets, word processors, or HTML).

Matching the Table Design to Predefined Specifications

Sometimes the end user of your tables has a definite format in mind, and no matter how hard it is to create, they refuse to be flexible. This example, and the others that follow, shows ways to create seemingly impossible tables.

For example, suppose your boss wants a table like the one in Figure 15.1. Instead of having a row heading for the N statistic, the table uses a column header in the format N=xx. It's easy to put an N in the column header, but how do you attach the value (number of observations)?

The answer is to cheat. You create a table cell that looks like a column header. Actually it is a table cell that displays a value, and uses a special format to generate the 'N=' prefix.

As a first step, let's build the basic table, leaving the N label in the row header for now.

```
PROC TABULATE DATA=TEMP F=10.;
   CLASS GENDER EDUC;
   VAR INCOME;
   TABLE ALL=' '*N
      (GENDER=' ' ALL)*INCOME=' '*MEAN=' ',
      EDUC ALL / ROW=FLOAT BOX='Mean Income';
RUN;
```

In Output 15.1, you can see that this creates a table with the N for each column at the top of each column. All that is missing is the 'N='attached to each N value. To do that, you can use a PICTURE statement in PROC FORMAT to generate a format with the 'N=' prefix.

```
PROC FORMAT;
   PICTURE NPIC 0-HIGH='0000' (PREFIX='N=');
RUN;
```

Now we can apply this format to the N value, and remove the N label from the row heading. The following code generates the table shown in Output 15.2.

```
PROC TABULATE DATA=TEMP F=10.;
   CLASS GENDER EDUC;
   VAR INCOME;
   TABLE ALL=' '*N=' '*F=NPIC.
      (GENDER=' ' ALL)*INCOME=' '*MEAN=' ',
      EDUC ALL / ROW=FLOAT BOX='Mean Income';
RUN;
```

For another way to add N= labels, see Tien et al (1997) in the "References" section.

Figure 15.1

Mean Income	Education			ALL
	<HS	HS	College	
	N=xx	N=yy	N=zz	N=xyz
Male				
Female				
ALL				

Output 15.1

```
----------------------------------------------------------------------
|Mean Income                  |          Education          |         |
|                             |-----------------------------|         |
|                             |  <HS   |   HS   | College |   ALL     |
|-----------------------------+--------+--------+---------+---------- |
|                      |N     |    1001|    2282|    3356 |    6639   |
|-----------------------------+--------+--------+---------+---------- |
|Male                         |   15673|   26621|   42977 |   33485   |
|-----------------------------+--------+--------+---------+---------- |
|Female                       |    8862|   13760|   23412 |   17780   |
|-----------------------------+--------+--------+---------+---------- |
|ALL                          |   11938|   19582|   32711 |   25066   |
----------------------------------------------------------------------
```

Output 15.2

```
----------------------------------------------------------------------
|Mean Income                  |          Education          |         |
|                             |-----------------------------|         |
|                             |  <HS   |   HS   | College |   ALL     |
|-----------------------------+--------+--------+---------+---------- |
|                             | N=1001 | N=2282 |  N=3356 |  N=6639   |
|-----------------------------+--------+--------+---------+---------- |
|Male                         |   15673|   26621|   42977 |   33485   |
|-----------------------------+--------+--------+---------+---------- |
|Female                       |    8862|   13760|   23412 |   17780   |
|-----------------------------+--------+--------+---------+---------- |
|ALL                          |   11938|   19582|   32711 |   25066   |
----------------------------------------------------------------------
```

Renaming the Statistics and Variables

While it's easy to build a table that displays the required results, the trick is to make it look nice. PROC TABULATE is a powerful tool, but it lacks a sense of aesthetics. Often you will need to get creative with your syntax to create a more elegant table.

For example, in the output shown in Output 15.3, you can see that while this table does show the N and MEAN by gender for each of the row variables, it looks pretty ugly. It turns out that the N in each column is the same for every row, so there's a lot of repetitive information in the table.

We can get rid of the redundant information by showing the N for just one of the row variables and moving the statistics to the row dimension of the table. This output, shown in Output 15.4, looks a lot better. Now there is no wasted information in the table.

```
PROC TABULATE DATA=TEMP;
   CLASS GENDER;
   VAR AGE INCOME NUMKIDS YRSEDUC;
   TABLE AGE=' '*N*F=8.
      (AGE INCOME NUMKIDS YRSEDUC)*MEAN*F=8.2,
      GENDER / ROW=FLOAT;
RUN;
```

But we can go one step further. To get rid of the repeated MEAN label in the rows, we can cheat. Because the one row that shows an N is already labeled as such, we can move the MEAN label to the column dimension. However, we can't move the MEAN *statistic* to the column dimension because all statistics have to be in the same dimension in a TABLE statement. If we move the MEAN, we'd have to move the N, and then we'd be back where we started.

Instead, we can move the MEAN *label* to the column dimension. First, the MEAN statistic is set to a null label. Then, the GENDER variable label is changed to read "MEAN." We can do this because the GENDER label is not needed, as the CLASS values of 'Male' and 'Female' are self-explanatory. A format is used to put parentheses around the N values so that they are distinct from the other values:

```
PROC FORMAT;
   PICTURE PARENFT LOW-HIGH='00000)' (PREFIX='(');
RUN;

PROC TABULATE DATA=TEMP;
   CLASS GENDER;
   VAR AGE INCOME NUMKIDS YRSEDUC;
   TABLE AGE=' '*N='(N)'*F=PARENFT.
      (AGE INCOME NUMKIDS YRSEDUC)*MEAN=' '*F=8.2,
      GENDER='MEAN' / ROW=FLOAT;
RUN;
```

The result is Output 15.5: a compact, elegant table that shows just the necessary information. Whenever you have a table that just doesn't look good, think creatively about moving or renaming the statistics to clean it up.

Output 15.3

```
----------------------------------------------------------------
|                               |          Gender              |
|                               |------------------------------|
|                               |     Male      |    Female    |
|                               |---------------+--------------|
|                               |  N  |  MEAN   |  N  |  MEAN  |
|-------------------------------+-----+---------+-----+--------|
|Age                            | 2980|  47.74  | 3432|  49.73 |
|-------------------------------+-----+---------+-----+--------|
|Income                         | 2980|33659.02 | 3432|17891.99|
|-------------------------------+-----+---------+-----+--------|
|No. of Children                | 2980|   1.49  | 3432|   1.48 |
|-------------------------------+-----+---------+-----+--------|
|Yrs Educ                       | 2980|  13.39  | 3432|  13.16 |
----------------------------------------------------------------
```

Output 15.4

```
-------------------------------------------------------
|                             |      Gender          |
|                             |----------------------|
|                             |   Male  |  Female    |
|-----------------------------+---------+------------|
|Age            |N            |   2980  |    3432    |
|---------------+-------------+---------+------------|
|Age            |MEAN         |  47.74  |   49.73    |
|---------------+-------------+---------+------------|
|Income         |MEAN         |33659.02 |  17891.99  |
|---------------+-------------+---------+------------|
|No. of Children|MEAN         |   1.49  |    1.48    |
|---------------+-------------+---------+------------|
|Yrs Educ       |MEAN         |  13.39  |   13.16    |
-------------------------------------------------------
```

Output 15.5

```
-------------------------------------------------------
|                             |       MEAN           |
|                             |----------------------|
|                             |   Male   |  Female   |
|-----------------------------+----------+-----------|
|(N)                          |  (2980)  |  (3432)   |
|-----------------------------+----------+-----------|
|Age                          |   47.74  |   49.73   |
|-----------------------------+----------+-----------|
|Income                       | 33659.02 | 17891.99  |
|-----------------------------+----------+-----------|
|No. of Children              |    1.49  |    1.48   |
|-----------------------------+----------+-----------|
|Yrs Educ                     |   13.39  |   13.16   |
-------------------------------------------------------
```

Calculating Percentages Another Way

Sometimes PROC TABULATE gives you more output than you need. For example, if you are trying to produce a table that shows what percent of your population is female, you can easily get that number by calculating the PCTN statistic for your gender variable. However, this not only gives you the percentage of females but it also gives you the percentage of males.

The following code produces a table that shows the N and PCTN for females (as well as the N and PCTN for males):

```
PROC TABULATE DATA=TEMP F=8.;
   CLASS RACE GENDER;
   TABLE GENDER=' '*(N PCTN<GENDER>*F=8.1) ALL*N,
      RACE / ROW=FLOAT;
RUN;
```

In the table shown in Output 15.6, you can see that much of the table is redundant. If the population is 53.6% female, then it follows that the remaining 46.4% are male. If you could get rid of the N and PCTN rows for males, you could save space. But how do you get PROC TABULATE to report only the percent female?

The answer is that you have to cheat. Instead of computing the PCTN of gender, you create a dummy (0 or 1) variable called FEMALE. For records where GENDER='Female', you set FEMALE=1 for records where GENDER='Male', you set FEMALE=0. Then (this is the trick) you take the *mean* of FEMALE, which will give you the percent of the population that is female. To get the N for females, you take the *sum* of FEMALE.

To see how this works, imagine that you have a file with ten records. Of the ten, six are female and four are male. If you compute the PCTN for GENDER, you find out that your population is 60% female and 40% male. If you instead create the variable FEMALE, it is set to 1 for six observations and set to 0 for four observations. To get the percentage of females, you take the mean of FEMALE and you get 0.6 or 60%. To get the number of observations for females, you sum the variable FEMALE and get 6.

```
DATA TEMP2;
   SET TEMP;
   IF GENDER='Female' THEN FEMALE=1;
   ELSE IF GENDER='Male' THEN FEMALE=0;
RUN;

PROC TABULATE DATA=TEMP2 F=8.;
   CLASS RACE;
   VAR FEMALE;
   TABLE FEMALE*(SUM='N' MEAN='PCTN'*F=PERCENT8.1) ALL*N,
      RACE / ROW=FLOAT;
RUN;
```

In your PROC TABULATE code, all you have to do is compute the SUM and MEAN of FEMALE, and use labels to identify the statistics as 'N' and 'PCTN'. You can also use the PERCENTw.d format to make the table values appear as percentages. This format multiplies values by 100 and attaches a percent sign. This table is shown in Output 15.7.

Output 15.6

```
---------------------------------------------------------------
|               |               |              Race             |
|               |               |-------------------------------|
|               |               | White  | Black  | Other  |
|---------------+---------------+--------+--------+--------|
|Male           |N              |   2908 |     76 |      7 |
|               |---------------+--------+--------+--------|
|               |PCTN           |   46.4 |   46.1 |   46.7 |
|---------------+---------------+--------+--------+--------|
|Female         |N              |   3353 |     89 |      8 |
|               |---------------+--------+--------+--------|
|               |PCTN           |   53.6 |   53.9 |   53.3 |
|---------------+---------------+--------+--------+--------|
|ALL            |N              |   6261 |    165 |     15 |
---------------------------------------------------------------
```

Output 15.7

```
---------------------------------------------------------------
|               |               |              Race             |
|               |               |-------------------------------|
|               |               | White  | Black  | Other  |
|---------------+---------------+--------+--------+--------|
|FEMALE         |N              |   3353 |     89 |      8 |
|               |---------------+--------+--------+--------|
|               |PCTN           |  53.6% |  53.9% |  53.3% |
|---------------+---------------+--------+--------+--------|
|ALL            |N              |   6261 |    165 |     15 |
---------------------------------------------------------------
```

Be aware that you can only use the PERCENTw.d format on a PROC TABULATE percent if you have computed it in the manner of this example by using a MEAN. If you apply a PERCENTw.d format to a regular PROC TABULATE percent that was computed using PCTN or PCTSUM, you will get percentages like 4000%. This happens because the values produced by the PCTN and PCTSUM statistics have already been multiplied by 100. If you apply the PERCENTw.d format, you will multiply them by 100 again, so 40% turns into 4000%. If you want to attach a percent sign to a PCTN or PCTSUM value, see the example on formatting percentages in Chapter 11, "Formatting Table Values."

Displaying the Results of Multiple-Response Questions

This problem comes up a lot in survey research. Instead of having a question where the respondent picks one answer from multiple choices, the question asks the respondent to check all answers that apply. These answers are then stored in multiple variables, which makes them hard to display in meaningful tables.

The table shown in Output 15.8 illustrates the problem. It shows the answers to a five-item multiple-response question. The question was asked: "Why did you choose to shop in our store? (check all that apply)." In this format, you can't compare the response rates easily.

The following code shows how to reorganize the data so that you can create an easy to read table that displays the results of a multiple-response question in descending order of frequency.

```
PROC FORMAT;
    VALUE RESPFT  1='Prices'
                  2='Service'
                  3='Location'
                  4='Hours'
                  5='Variety';
RUN;
DATA TEMPANS2;
   SET TEMPANS;
   ARRAY ANS{5} ANS1-ANS5;
   IF SUM(ANS1,ANS2,ANS3,ANS4,ANS5)>0;
   DO A=1 TO 5;
      IF ANS{A} NE . THEN DO;
         RESP=A;
         ANSWER=ANS{A};
         OUTPUT;
      END;
   END;
RUN;
PROC SUMMARY DATA=TEMPANS2 NWAY;
   CLASS RESP;
   VAR ANSWER;
   OUTPUT OUT=TEMPANS3 MEAN=;
RUN;
PROC SORT DATA=TEMPANS3; BY DESCENDING ANSWER; RUN;
PROC TABULATE DATA=TEMPANS3 ORDER=DATA;
   CLASS RESP;
   VAR ANSWER;
   TABLE RESP=' ',
      ANSWER='Percent of respondents'*MEAN=' '*F=PERCENT14.1
      / RTS=25 BOX='Reason for choosing our store';
   FORMAT RESP RESPFT.;
RUN;
```

The output is shown in Output 15.9. Now it is easy to see that location and service are far more important than prices, an important result to share with management. You will notice that the results change slightly. This is because the revised table includes all people who answered at least one item. The original table could show only people who had responses to all five items.

Output 15.8

```
-------------------------------------------
|Reason for choosing our|  Responses   |
|store                  |    (Pct.)    |
|-----------------------+--------------|
|Prices                 |              |
|No                     |         97.7 |
|Yes                    |          2.3 |
|Service                |              |
|No                     |         46.4 |
|Yes                    |         53.6 |
|Location               |              |
|No                     |         25.7 |
|Yes                    |         74.3 |
|Hours                  |              |
|No                     |         59.2 |
|Yes                    |         40.8 |
|Variety                |              |
|No                     |         51.3 |
|Yes                    |         48.7 |
-------------------------------------------
```

Output 15.9

```
-------------------------------------------
|Reason for choosing our|  Percent of  |
|store                  |  respondents |
|-----------------------+--------------|
|Location               |        77.5% |
|-----------------------+--------------|
|Service                |        55.9% |
|-----------------------+--------------|
|Variety                |        50.7% |
|-----------------------+--------------|
|Hours                  |        42.6% |
|-----------------------+--------------|
|Prices                 |         2.4% |
-------------------------------------------
```

Using a Variable for Both Classification and Analysis

This issue also comes up a lot in survey research. Frequently you want to display the answer to a multiple-choice survey question both as the percentage of respondents who chose each category and also as a mean rating in the same table.

For example, the following code is designed to produce a table showing satisfaction scores by customer type and size. The code tries to build a table that shows both the percentage of responses in each satisfaction category as well as the mean satisfaction score.

```
PROC TABULATE DATA=TEMP;
   CLASS SAT CUSTTYPE SIZE;
   VAR SAT;
   TABLE CUSTTYPE=' '*SIZE=' ',
      SAT*PCTN<SAT>=' '*F=PCTPIC. SAT*MEAN*F=12.1;
RUN;
```

But when you run the code, you get the error message shown in Output 15.10. This is because the variable SAT is listed in both the CLASS statement and the VAR statement.

If you try to fix this code by removing the VAR statement, you'll get a new error message, as shown in Output 15.11. You can't compute the MEAN of a CLASS variable.

If you try to fix this code by removing SAT from the CLASS statement, you won't get an error message. Instead you'll get the nonsensical table shown in Output 15.12. Because SAT is not defined as a CLASS variable, the PCTN statistic does not work correctly.

To fix the problem, we need a way to list the SAT variable in both the CLASS and VAR statements. The solution is to make a copy of SAT. This way we can list the original variable in the CLASS statement and the copy in the VAR statement. Then we won't get an error message, and the table will work properly. The following code does the trick:

```
DATA TEMP2;
   SET TEMP;
   SATN=SAT;
RUN;
PROC TABULATE DATA=TEMP2;
   CLASS SAT CUSTTYPE SIZE;
   VAR SATN;
   TABLE CUSTTYPE=' '*SIZE=' ',
      SAT*PCTN<SAT>=' '*F=PCTPIC. SATN=' '*MEAN*F=8.1;
RUN;
```

As you can see from Output 15.13, the new table works perfectly. The variable SAT produces the correct percentages for each category, and the variable SATN is used to compute the MEAN.

Output 15.10

```
14   PROC TABULATE DATA=TEMP;
15        CLASS SAT CUSTTYPE SIZE;
16        VAR SAT;
17        TABLE CUSTTYPE=' '*SIZE=' ',
18             SAT*PCTN<SAT>=' '*F=PCTPIC. SAT*MEAN*F=12.1;
19   RUN;


ERROR: SAT appears in both CLASS and VAR list.
NOTE: The SAS System stopped processing this step because of errors.
```

Output 15.11

```
20   PROC TABULATE DATA=TEMP;
21        CLASS SAT CUSTTYPE SIZE;
22        TABLE CUSTTYPE=' '*SIZE=' ',
23             SAT*PCTN<SAT>=' '*F=PCTPIC. SAT*MEAN*F=12.1;|
24   RUN;


ERROR: Statistic other than N was requested without analysis variable in the following nesting :
CUSTTYPE * SIZE * SAT * MEAN * F.
NOTE: The SAS System stopped processing this step because of errors.
```

Output 15.12

```
-----------------------------------------------------------
|                             |          |Satisfaction |
|                             |Satisfa- |  Rating     |
|                             | ction   |-------------|
|                             | Rating  |    MEAN     |
|-----------------------------+---------+-------------|
|Retail        |Large         | 100.0%|          3.5 |
|              |--------------+---------+-------------|
|              |Small         | 100.0%|          3.6 |
|--------------+--------------+---------+-------------|
|Wholesale     |Large         | 100.0%|          3.6 |
|              |--------------+---------+-------------|
|              |Small         | 100.0%|          3.6 |
-----------------------------------------------------------
```

Output 15.13

```
------------------------------------------------------------------------------------
|                         |            Satisfaction Rating              |          |
|                         |---------------------------------------------|          |
|                         |   1   |   2   |   3   |   4   |   5   | MEAN  |
|-------------------------+-------+-------+-------+-------+-------+--------|
|Retail      |Large       | 10.6%| 12.5%| 18.9%| 28.7%| 29.1%|    3.5 |
|            |------------+-------+-------+-------+-------+-------+--------|
|            |Small       | 10.2%| 10.9%| 19.3%| 29.7%| 29.6%|    3.6 |
|------------+------------+-------+-------+-------+-------+-------+--------|
|Wholesale   |Large       | 10.4%|  9.9%| 19.8%| 30.4%| 29.2%|    3.6 |
|            |------------+-------+-------+-------+-------+-------+--------|
|            |Small       |  9.7%|  9.8%| 20.5%| 30.8%| 28.9%|    3.6 |
------------------------------------------------------------------------------------
```

Creating a One-Dimensional Table That Runs Vertically

When you build a one-dimensional table, PROC TABULATE takes your table definition and builds a table with a single row and many columns. There is no option to ask for a one-dimensional table that has a single column and many rows.

This can be a problem if your table turns out to be extremely wide. When your table becomes too wide for the page, PROC TABULATE generates a "Continued" message and builds the additional columns on a second page. This makes your table hard to read.

For example, take the following table. This shows the number of observations in the data set for each occupation:

```
PROC TABULATE DATA=TEMP;
   CLASS OCCUP;
   TABLE OCCUP=' '*N*F=15.;
RUN;
```

In the output shown in Output 15.14, you can see that this table is so wide that it has to be continued on a second page.

Part of the reason the table is so wide is that the columns have to be fairly wide to hold the long labels. You could shorten or hyphenate the labels and squash the table onto a single page, but wouldn't it look better if the table could be a one-dimensional table that runs vertically? Then the occupation descriptions would be row headings, and it would be easy to allow enough space for the text.

Another solution would be to print the table in landscape orientation. This would allow more columns to fit on the page. But whenever possible, it's best to avoid producing tables that have to be printed in landscape orientation. It doesn't matter whether you're producing a market analysis for a client or an internal financial report for your boss. The people who have to read your table will appreciate it if they don't have to turn the page around to go from reading the text to reading a table and back to the text again.

So the best solution would be to turn the table instead of the page. If the table ran vertically instead of horizontally, it would fit on the page. PROC TABULATE doesn't have an option to print a one-dimensional table that has rows but not columns. However, you can trick PROC TABULATE into thinking that a one-dimensional table is a two-dimensional table, even if you only have one variable. You do this with the ALL keyword:

```
PROC TABULATE DATA=TEMP;
   CLASS OCCUP;
   TABLE OCCUP=' '*N=' '*F=8.,
      ALL='N'
      / ROW=FLOAT;
RUN;
```

In this example, the TABLE statement is left unchanged, except for adding ALL as a column dimension. This forces PROC TABULATE to treat OCCUP as a row variable. Because ALL merely selects all observations and reports the selected statistic for all of the observations, it has no effect except to create a second dimension for the table. The only other change made to the code was to move the 'N' label to the column header because a header labeled 'ALL' would be confusing. The resulting table is shown in Output 15.15.

Output 15.14

```
---------------------------------------------------------------------
| Managerial   | Professional  |  Technical   |    Sales     |
|--------------+---------------+--------------+--------------|
|     N        |      N        |      N       |      N       |
|--------------+---------------+--------------+--------------|
|         770|           865|          144|          447|
---------------------------------------------------------------------
```

(CONTINUED)

```
---------------------------------------------------------------------
|  Clerical    |   Services   | Manufacturing |   Farming    |
|--------------+--------------+---------------+--------------|
|     N        |      N       |      N        |      N       |
|--------------+--------------+---------------+--------------|
|         676|          530|          733|           79|
---------------------------------------------------------------------
```

Output 15.15

```
-----------------------------
|                 |    N    |
|-----------------+-------- |
|Managerial       |    770  |
|-----------------+-------- |
|Professional     |    865  |
|-----------------+-------- |
|Technical        |    144  |
|-----------------+-------- |
|Sales            |    447  |
|-----------------+-------- |
|Clerical         |    676  |
|-----------------+-------- |
|Services         |    530  |
|-----------------+-------- |
|Manufacturing    |    733  |
|-----------------+-------- |
|Farming          |     79  |
-----------------------------
```

Handling CLASS Variables with Missing Data

If you work with survey data, this problem will be familiar. Let's say you have a questionnaire where each question is answered on a scale of 1 to 5. When you look at the data, you discover that for one of the questions, no one ever selected answer 2.

If you produce a PROC TABULATE table to show the questionnaire answers broken down by gender, you discover that PROC TABULATE only displays the categories of 1, 3, 4, and 5 for the question. In the table shown in Output 15.16, you can see that the category 2 is dropped from the table. It would be easier to read the table if you could force PROC TABULATE to display all of the valid codes even if they have no observations.

Unfortunately, PROC TABULATE will not do this. Any time a category of a CLASS variable has no observations, that category is dropped from the table. The following code shows a way to get around this. Be aware, though, that this is a *sneaky trick* and needs to be *well documented* in your code.

The trick is to create some fake data with no missing values for any category of the CLASS variables. The fake data is used to create the second page of a three-dimensional table. Then, because it's a three-dimensional table, the PRINTMISS option can be used. This forces both pages of the table to show all categories of CLASS variables that have valid data on at least one of the two pages.

```
DATA FIXIT;
   DO A=1 TO 5;
      DO S=1 TO 2;
         Q1=A; GENDER=S; OUTPUT;
      END;
   END;
RUN;

DATA FIXED;
   SET TEMP (IN=REAL) FIXIT (IN=FAKE);
   IF REAL THEN KEEPIT='Table with no missing data';
   ELSE IF FAKE THEN KEEPIT='FAKE DATA - DO NOT USE';
RUN;

PROC TABULATE DATA=FIXED;
   CLASS KEEPIT GENDER Q1;
   TABLE KEEPIT=' ', Q1*N=' '*F=8., GENDER ALL /
      BOX=_PAGE_ PRINTMISS ROW=FLOAT;
RUN;
```

To create the fake data, this code uses nested DO loops to go through all possible combinations of the CLASS variables and create a single record for each. This ensures that the second page of the new table, which uses the fake data, will have one observation in every cell. A variable called KEEPIT is used to keep track of which data is the fake data and which is the real data. KEEPIT is then used as the page variable in the TABLE statement.

The resulting output is shown in Output 15.17. All you have to do to use this code is be sure to throw out the table labeled 'FAKE DATA - DO NOT USE'.

Output 15.16

```
--------------------------------------------------------------
|                              |      Gender     |          |
|                              |-----------------|          |
|                              | Male  | Female |  ALL     |
|------------------------------+-------+--------+--------- |
|Q1                            |       |        |          |
|------------------------------|       |        |          |
|1                             |     4 |      7 |     11   |
|------------------------------+-------+--------+--------- |
|3                             |    20 |     23 |     43   |
|------------------------------+-------+--------+--------- |
|4                             |    45 |     50 |     95   |
|------------------------------+-------+--------+--------- |
|5                             |    58 |     77 |    135   |
--------------------------------------------------------------
```

Output 15.17

```
--------------------------------------------------------------
|FAKE DATA - DO NOT USE        |      Gender     |          |
|                              |-----------------|          |
|                              | Male  | Female |  ALL     |
|------------------------------+-------+--------+--------- |
|Q1                            |       |        |          |
|------------------------------|       |        |          |
|1                             |     1 |      1 |      2   |
|------------------------------+-------+--------+--------- |
|2                             |     1 |      1 |      2   |
|------------------------------+-------+--------+--------- |
|3                             |     1 |      1 |      2   |
|------------------------------+-------+--------+--------- |
|4                             |     1 |      1 |      2   |
|------------------------------+-------+--------+--------- |
|5                             |     1 |      1 |      2   |
--------------------------------------------------------------


--------------------------------------------------------------
|Table with no missing data    |      Gender     |          |
|                              |-----------------|          |
|                              | Male  | Female |  ALL     |
|------------------------------+-------+--------+--------- |
|Q1                            |       |        |          |
|------------------------------|       |        |          |
|1                             |     4 |      7 |     11   |
|------------------------------+-------+--------+--------- |
|2                             |     . |      . |      .   |
|------------------------------+-------+--------+--------- |
|3                             |    20 |     23 |     43   |
|------------------------------+-------+--------+--------- |
|4                             |    45 |     50 |     95   |
|------------------------------+-------+--------+--------- |
|5                             |    58 |     77 |    135   |
--------------------------------------------------------------
```

In Version 7, you can use the PRELOADFMT and EXCLUSIVE options to get around this problem. See Chapter 24, "Version 7 Enhancements," for more information.

Creating Footnote References in Table Cells - I

Many times when you build a table, you want to add footnotes to explain something about the underlying data. If the footnote applies to the entire table, it is easy to add a footnote below the table by using a FOOTNOTE statement. For example, in Output 15.18, the first footnote tells readers that the data comes from the XYZ Corporation Survey.

But what if a footnote applies only to a single row or column of the table? An easy solution is to add an asterisk to the appropriate row or column label, and then use the asterisk in the footnote label. This code produces the table shown in Output 15.18. It has a footnote reference in the column heading created by changing the variable's label, and a footnote reference in a row heading created by changing the variable's format.

```
PROC FORMAT;
    VALUE FOOTFT  ... other values skipped to save space ...
                  11='Farming**';
RUN;
FOOTNOTE 'Source: XYZ Company Survey.';
FOOTNOTE2 '* Part-time includes employees working <=30 hrs/wk.';
FOOTNOTE3 '** Does not include seasonal employees.';
PROC TABULATE DATA=TEMP;
    CLASS OCCUP FULLTIME;
    FORMAT OCCUP FOOTFT.;
    TABLE OCCUP=' '*N=' ',
        FULLTIME='Employment Status*'*F=10.
        / BOX='Number of Employees' ROW=FLOAT;
RUN;
```

But what if you need to reference a single cell in the table, instead of an entire row or column? The solution is to trick PROC TABULATE by using a PICTURE format. First you run the code and determine what value will end up in the table cell where you want to add a footnote reference. Then you assign a special format with an asterisk for that particular value. Then you run the code again and the asterisk will be added to the table cell.

```
PROC FORMAT;
    PICTURE FOOTPIC 0-18,20-HIGH='0000000000'
                    19='00000000**';
RUN;
PROC TABULATE DATA=TEMP;
    CLASS OCCUP FULLTIME;
    TABLE OCCUP=' '*N=' ',
        FULLTIME='Employment Status*'*F=FOOTPIC.
        / BOX='Number of Employees' ROW=FLOAT;
RUN;
```

The resulting table is shown in Output 15.19. However, you have to be extremely careful when using this technique. If the underlying data changes, causing the value in your targeted cell to change, the footnote will disappear from that cell. And it may even appear in an unintended cell if that value pops up somewhere else. Also, this approach will not work if the cell where you want to put the footnote does not have a unique value.

Output 15.18

```
-------------------------------------------------------------
|Number of Employees            | Employment Status*  |
|                               |--------------------|
|                               |Full time |Part time |
|-------------------------------+----------+--------- |
|Managerial                     |      670 |       74 |
|-------------------------------+----------+--------- |
|Professional                   |      745 |       89 |
|-------------------------------+----------+--------- |
|Technical                      |      118 |       16 |
|-------------------------------+----------+--------- |
|Sales                          |      373 |       39 |
|-------------------------------+----------+--------- |
|Clerical                       |      562 |       74 |
|-------------------------------+----------+--------- |
|Services                       |      439 |       47 |
|-------------------------------+----------+--------- |
|Manufacturing                  |      567 |       92 |
|-------------------------------+----------+--------- |
|Farming**                      |       51 |       19 |
-------------------------------------------------------------
```

Source: XYZ Company Survey.
* Part-time includes employees working <30 hrs/wk.
** Does not include seasonal employees.

Output 15.19

```
-------------------------------------------------------------
|Number of Employees            | Employment Status*  |
|                               |--------------------|
|                               |Full time |Part time |
|-------------------------------+----------+--------- |
|Managerial                     |      670 |       74 |
|-------------------------------+----------+--------- |
|Professional                   |      745 |       89 |
|-------------------------------+----------+--------- |
|Technical                      |      118 |       16 |
|-------------------------------+----------+--------- |
|Sales                          |      373 |       39 |
|-------------------------------+----------+--------- |
|Clerical                       |      562 |       74 |
|-------------------------------+----------+--------- |
|Services                       |      439 |       47 |
|-------------------------------+----------+--------- |
|Manufacturing                  |      567 |       92 |
|-------------------------------+----------+--------- |
|Farming                        |       51 |     19** |
-------------------------------------------------------------
```

Source: XYZ Company Survey.
* Part-time includes employees working <30 hrs/wk.
** Does not include seasonal employees.

Creating Footnote References in Table Cells - II

The previous example showed how to add footnote references to explain a particular row, column, or cell in the data. Another way you can use footnotes is to identify a particular range of values. For example, you might want to identify all cells where income is less than $20,000 or where it is greater than $100,000.

To assign a footnote reference to a range of values, you can use a PICTURE format. In this first example, all values of $20,000 or more are formatted as integers. The integers are given a width of 10 so that the columns will be wide enough to fit the headings. But values under $20,000 are displayed as an asterisk, referencing the footnote '<$20,000'.

```
PROC FORMAT;
   PICTURE LOWNOTE LOW-<20000='*'
                   20000-HIGH='0000000000';
RUN;
FOOTNOTE '* <$20,000';
PROC TABULATE DATA=TEMP;
   CLASS OCCUP FULLTIME;
   VAR INCOME;
   TABLE OCCUP=' '*INCOME=' '*MEAN=' '*F=LOWNOTE.,
      FULLTIME=' '
      / BOX='Mean Income' ROW=FLOAT ;
RUN;
```

In the table shown in Output 15.20, two cells have a mean income less than $20,000. These cells are marked with an asterisk.

In the second example, all values up to $100,000 are formatted as integers. But values over $100,000 are displayed as an asterisk, referencing the footnote '>$100,000'.

```
PROC FORMAT;
   PICTURE HIGHNOTE   LOW-100000='0000000000'
                      100001-HIGH='*';
RUN;
FOOTNOTE '* >$100,000';
PROC TABULATE DATA=TEMP;
   CLASS OCCUP FULLTIME;
   VAR INCOME;
   TABLE OCCUP=' '*INCOME=' '*MEAN=' '*F=HIGHNOTE.,
      FULLTIME=' '
      / BOX='Mean Income' ROW=FLOAT ;
RUN;
```

In Output 15.21, the table shows that one cell has a mean income greater than $100,000. This cell is marked with an asterisk.

Output 15.20

Mean Income	Full time	Part time
Managerial	189202	37303
Professional	40807	41894
Technical	31028	31722
Sales	33023	27997
Clerical	22656	24766
Services	*	21237
Manufacturing	28776	23417
Farming	*	22731

* <$20,000

Output 15.21

Mean Income	Full time	Part time
Managerial	*	37303
Professional	40807	41894
Technical	31028	31722
Sales	33023	27997
Clerical	22656	24766
Services	18270	21237
Manufacturing	28776	23417
Farming	17952	22731

* >$100,000

Creating Footnotes to Show Additional Statistical Results

If you want to test whether the means displayed in your table are significantly different from zero, you're in luck. This is the one statistical test that PROC TABULATE knows how to do. A table with means and t-tests is shown in Output 15.22. The code for this table is as follows:

```
FOOTNOTE '* P-VALUE for test if mean is signif. different from 0';
PROC TABULATE DATA=TEMP;
   CLASS RACE;
   VAR CHANGEWT;
   TABLE RACE=' ',
      CHANGEWT*(MEAN PRT='P-Value*')*F=8.4;
RUN;
```

Unfortunately, this is the only statistical test that PROC TABULATE can run. And it's pretty rare that this is the statistical test that you want. When you're doing a cross-tabulation, the statistic that's most likely to come up is a chi-square test. If you want to find out if the frequency distribution in a table is statistically significant, this is the test that you want.

Luckily for you, there is another way to attach statistical test results to a PROC TABULATE table. All you have to do is run your statistical procedure, output the results to a data set, and then use a macro variable to add the results to your table as a footnote. For example, say you've created a table that shows employment status by gender. From looking at Output 15.23, it appears as if more part-time jobs are held by women. To test this theory, you do a chi-square test on the table using PROC FREQ. The following code shows how to take the test result and add it to the table as a footnote.

```
PROC FREQ DATA=TEMP NOPRINT;
   OUTPUT OUT=CHI PCHI;
   TABLES GENDER*FULLTIME / CHISQ;
RUN;
DATA _NULL_;
   SET CHI;
   CALL SYMPUT('CHI',PUT(P_PCHI,8.4));
RUN;
FOOTNOTE "P-value for chi-square test = &CHI";
PROC TABULATE DATA=TEMP;
   CLASS GENDER FULLTIME;
   TABLE GENDER=' ', FULLTIME=' '*N=' '*F=12.;
RUN;
```

The PROC FREQ output is directed to the data set CHI. A printout of this data set is shown in Output 15.24. A DATA step is then used to read in the results, and the CALL SYMPUT macro function is used to assign the p-value for the chi-square test to the macro variable CHI. Finally, the macro variable CHI is used in the FOOTNOTE definition to display the p-value at the bottom of the table. The resulting table is shown in Output 15.25.

This approach will work for any statistic that can be output to a data set. Most SAS statistical procedures support output data sets, though not all statistics are available in these output data sets.

Output 15.22

```
-----------------------------------------------------
|                                  |Change in Weight |
|                                  |-----------------|
|                                  |  MEAN  |P-value* |
|----------------------------------+--------+--------|
|White                             |  0.0050|  0.0001|
|----------------------------------+--------+--------|
|Black                             |  0.0004|  0.3580|
|----------------------------------+--------+--------|
|Other                             |  0.0054|  0.0001|
-----------------------------------------------------
```

* P-value for test if mean is signif. different from 0.

Output 15.23

```
-----------------------------------------------------------
|                               | Full time | Part time  |
|-------------------------------+-----------+-----------|
|Male                           |      1978|        221 |
|-------------------------------+-----------+-----------|
|Female                         |      1787|        247 |
-----------------------------------------------------------
```

Output 15.24

```
    OBS      _PCHI_      DF_PCHI      P_PCHI

     1      4.70950         1        0.029996
```

Output 15.25

```
-----------------------------------------------------------
|                               | Full time | Part time  |
|-------------------------------+-----------+-----------|
|Male                           |      1978|        221 |
|-------------------------------+-----------+-----------|
|Female                         |      1787|        247 |
-----------------------------------------------------------
```
* P-value for chi-square test =   0.0300

Version 7 expands your ability to combine statistics from other procedures with your PROC TABULATE output. Under Version 7, all statistical procedures produce output data sets, and the output data sets contain all statistics computed by the procedure.

Outputting the Results

If you're producing a table for your own benefit, you may not need a fancy report. You may be satisfied with a quick printout directly from SAS.

But if you will be sharing your results with others, chances are that you may need to bring your results into some other software package for presentation.

This chapter demonstrates how to export PROC TABULATE tables to a spreadsheet or word processor. This allows you to integrate your PROC TABULATE output with the rest of the information you are presenting into a single report.

These days, many reports are delivered via the Internet. This chapter also shows how to turn your PROC TABULATE tables into HTML-formatted output.

Exporting Tables to a Spreadsheet: Creating the File

There are two steps involved in moving your PROC TABULATE output from SAS into a spreadsheet. First, you have to create an output file that a spreadsheet can understand.

For example, take a look at the basic PROC TABULATE table shown in Output 16.1. This output, with its dashes, plus signs, and vertical bars for dividers, is easily understood when you view it in the output window or print it out. But to a spreadsheet application, it might as well be written in Greek. To export this table to a spreadsheet, we need to make a few modifications.

First, we need to send the output to a file. There are many ways to do this. You could cut and paste the table from the SAS output window or from your output file. But a better way to do this is to add a PROC PRINTTO to your code so that the exact output you want goes to a separate output file.

Another change we need to make is to increase the size of the page. We don't want the page breaks that SAS generates in the middle of our table. If we need to put page breaks into the table, it's better to do this later in the spreadsheet application. A LINESIZE and PAGESIZE of 250 should be big enough to hold the largest of tables.

Finally, we need to make two changes to the PROC TABULATE statement itself. First, we need to turn on the NOSEPS option. Normally, PROC TABULATE output is made up of two lines of text for each row in the table. The first line of text holds the row values, the second line of text holds the row dividers. This is confusing to a spreadsheet program because it doesn't know what to do with the dividers. Because spreadsheet rows already have dividers built in, we can dispense with the row separators.

The other change we need to make to the PROC TABULATE statement is to modify the FORMCHAR statement. We don't want to confuse a spreadsheet program with all of those dashes, plus signs, and vertical bars. We can re-create whatever borders we want for our table after it is imported into the spreadsheet. So the new FORMCHAR setting is all blank spaces except for the first character, which is a comma.

```
FILENAME FORXPORT 'C:\TEMP\TABULATE.LIS';
PROC PRINTTO PRINT=FORXPORT; RUN;

    OPTIONS LS=250 PS=250;
    PROC TABULATE DATA=TEMP NOSEPS FORMCHAR=',               ';
        CLASS OCCUP UNION GENDER;
        VAR EMPYEARS;
        TABLE UNION=' '*(GENDER=' ' ALL),
            OCCUP=' '*EMPYEARS=' '*MEAN=' '*F=12.2
            / BOX='Average years employment' ROW=FLOAT RTS=32;
    RUN;

    PROC PRINTTO; RUN;
```

If you look at the resulting file in Output 16.2, you can see that this FORMCHAR setting has created a text file where each of the columns in the table is delineated with a comma. So what we have created is a comma-delimited text file that is easily understood by a wide variety of spreadsheet programs.

Now we are ready to for the second step, importing this file into the spreadsheet application.

Output 16.1

```
----------------------------------------------------------------------------
|Average years employment      | Managerial |Professional| Technical |  Sales    |
|------------------------------+------------+------------+-----------+---------- |
|Non-Union    |Female          |       7.36|        6.19|       5.83|       4.81 |
|             |----------------+------------+------------+-----------+---------- |
|             |Male            |       6.56|        5.29|       4.18|       3.48 |
|             |----------------+------------+------------+-----------+---------- |
|             |ALL             |       6.93|        5.71|       4.96|       4.27 |
|-------------+----------------+------------+------------+-----------+---------- |
|Union        |Female          |       7.61|        5.36|       6.40|       3.47 |
|             |----------------+------------+------------+-----------+---------- |
|             |Male            |       5.31|        4.00|       4.68|       2.13 |
|             |----------------+------------+------------+-----------+---------- |
|             |ALL             |       6.58|        4.55|       5.69|       3.06 |
----------------------------------------------------------------------------
```

Output 16.2

```
,Average years employment      , Managerial ,Professional, Technical  ,   Sales     ,

,Non-Union    Female           ,       7.36,        6.19,       5.83,       4.81,
,             Male             ,       6.56,        5.29,       4.18,       3.48,
,             ALL              ,       6.93,        5.71,       4.96,       4.27,
,Union        Female           ,       7.61,        5.36,       6.40,       3.47,
,             Male             ,       5.31,        4.00,       4.68,       2.13,
,             ALL              ,       6.58,        4.55,       5.69,       3.06,
```

This output file is comma-delimited. However, sometimes you need to have commas in your table. Perhaps one of your variables has a comma in its label, or one of the class value labels contains a comma. Or maybe you want to use a format like DOLLAR that contains commas. This will cause problems when you import the file into the spreadsheet.

To get around this, pick a different delimiter. You can use a FORMCHAR='~   ' to create a file that is delimited by the tilde character. This character is unlikely to occur elsewhere in your table. Keep in mind is that when you import the file into the spreadsheet, you need to specify that a tilde is used as the delimiter.

In Version 7, you can create HTML output that can be imported directly into most spreadsheet applications. See Chapter 24, "Version 7 Enhancements," for more information.

Exporting Tables to a Spreadsheet: Importing the Files

After you've built the output file, the rest of this process is fairly simple. To get the output into the spreadsheet, select Open or Import in your spreadsheet application, identify the file as comma-delimited, and watch your table reappear as a spreadsheet. You may have to do a little reformatting of your headings, but the table values should be ready to go.

To illustrate this process, this example shows how to import a file into Microsoft Excel.[1] The process is very similar in other spreadsheet applications.

The first step is to select File and then Open from the main menu bar. This will bring up a dialog box where you can select your SAS output file. Excel will try to open the file, eventually figuring out that this is not an Excel file. At this point, you will be warned that this is not an Excel file and asked if you wish to continue anyway.

Display 16.1

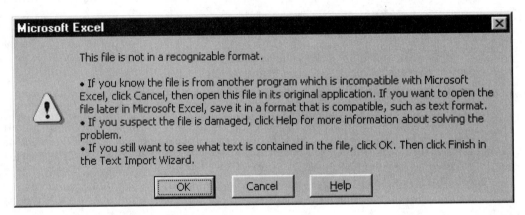

After you select OK, Excel launches the Text Import Wizard. This wizard walks you through the rest of the process of importing your file.

---

[1]   This example was run using Microsoft Excel 97. Import features are fairly similar in other versions of Excel and also in other spreadsheet software packages.

Display 16.2

The first step is to identify the type of file. Your options are delimited and fixed width. Because our file was created with delimiters, we will select delimited.

The only other option you might want to use on this page is the Start import at row: option. If there is other output in your file, or if you want to skip importing the headers, you can skip over rows by increasing this setting from its default of 1.

Display 16.3

The next step is to specify the type of delimiter. The next window gives you the choice of a number of different delimiters.

Because we used commas as our delimiters, we can check the box for comma.

Now we have our basic table moved into Excel. If you look closely at the spreadsheet, you can see that the table values look great, with each of the rows and columns is perfectly aligned. However, the row and column headings look a bit odd.

Excel has made every column the same width, but we know that our row headings need more room. Also, Excel has inserted an extra column to the left of the table.

Display 16.4

	A	B	C	D	E	F	G
1	☐The SAS System						
2							
3							
4		Average ye	Manageria	Profession	Technical	Sales	
5							
6		Non-Union	7.36	6.19	5.83	4.81	
7			6.56	5.29	4.18	3.48	
8			6.93	5.71	4.96	4.27	
9		Union	7.61	5.36	6.4	3.47	
10			5.31	4	4.68	2.13	
11			6.58	4.55	5.69	3.06	
12							

We can fix this by deleting the first column and widening the second column.

This looks better, but even after fixing the column width, the row headings still don't look right. This is because Excel has chosen its default font of Arial, which is a proportional font.

This means that skinny characters like the letter I are given less space than fat characters like the letter W. This changes the width of the labels in the row headings and makes the labels fail to line up.

Display 16.5

	A	B	C	D	E	F
1						
2						
3						
4	Average years employment	Manageria	Profession	Technical	Sales	
5						
6	Non-Union FEMALE	7.36	6.19	5.83	4.81	
7	MALE	6.56	5.29	4.18	3.48	
8	ALL	6.93	5.71	4.96	4.27	
9	Union FEMALE	7.61	5.36	6.4	3.47	
10	MALE	5.31	4	4.68	2.13	
11	ALL	6.58	4.55	5.69	3.06	
12						
13						

Display 16.6

To fix the row headings, we need to select a nonproportional font for our table.

The nonproportional fonts you have available will depend on your system, but a couple of good choices are Courier (which is on most systems) or SAS Monospace (which ships with SAS).

With the new font, the table looks much better. It looks just like the SAS table we started with.

	A	B	C	D	E	F
1						
2						
3						
4	Average years employment	Managerial	Professional	Technical	Sales	
5						
6	Non-Union FEMALE	7.36	6.19	5.83	4.81	
7	MALE	6.56	5.29	4.18	3.48	
8	ALL	6.93	5.71	4.96	4.27	
9	Union FEMALE	7.61	5.36	6.4	3.47	
10	MALE	5.31	4	4.68	2.13	
11	ALL	6.58	4.55	5.69	3.06	
12						

Display 16.7

Now we can start using the power of the spreadsheet application to improve and enhance the table.

With a few tweaks of cell borders, fonts and graphics, you can end up with something like the table shown in Display 16.7.

	A	B	C	D	E	F
1		**Average years employment**				
2						
3		Managerial	Professional	Technical	Sales	
4	Non-Union FEMALE	7.36	6.19	5.83	4.81	
5	MALE	6.56	5.29	4.18	3.48	
6	ALL	6.93	5.71	4.96	4.27	
7	Union FEMALE	7.61	5.36	6.40	3.47	
8	MALE	5.31	4.00	4.68	2.13	
9	ALL	6.58	4.55	5.69	3.06	
10						

The only limitation to importing tables into a spreadsheet is that complex tables with multiple column headings may need to have the headings adjusted. All of the data columns will line up, but the column headings for a group of columns will appear over the left-most column in the group. To fix this, center the column headings over the group of columns after importing the table into your spreadsheet. (This problem does not affect multiple row headings.)

Exporting Tables to a Word Processor

Exporting your PROC TABULATE output to a word processor is very similar to exporting to a spreadsheet. You prepare the output file in exactly the same way.

In fact, may of the steps in importing the data are also the same. This example will show how to import PROC TABULATE tables into Microsoft Word.[2] Importing your tables into other word processors will be very similar.

Display 16.8

The first step is to open your Word document. Then, with the cursor placed where you want the table to go, select `Insert` and then `File`.

This will bring up an Open File dialog where you can select your PROC TABULATE output file.

Display 16.9

The file will be brought into Word as text.

This file looks pretty much like it did in the SAS output window, except that it is too wide for the page, so the rows are wrapping onto a second line.

To turn this output back into a table, we need to tell Word that this text should be a table.

---

[2] This example was run using Microsoft Word 97. Import features are fairly similar in other versions of Word and also in other word processors.

Display 16.10

To convert the text to a table, you need to highlight the text that you just imported, and select Table and then Convert Text to Table from the menu bar.

This brings up a dialog box so that you can tell Word how your table is set up.

Under Separate text at, select Commas. Then you can click OK and let Word build your table.

Display 16.11

Don't be alarmed when you first see the table. Word is not as good as Excel at interpreting your output. The table will not look right at first.

As you can see in the output to the right, there are two extra columns in the table, one at the left and one at the right.

Because the two extra columns are blank, we can delete them. To delete a column, select the column by highlighting it or by using the `Select Column` command on the `Table` menu, and then select the `Delete Column` command on the `Table` menu. You can also delete the extra row in the same manner.

Display 16.12

	Average years employment	Manageri al	Professi onal	Technica l	Sales	
	Non-Union FEMALE	7.36	6.19	5.83	4.81	
	MALE	6.56	5.29	4.18	3.48	
	ALL	6.93	5.71	4.96	4.27	
	Union FEMALE	7.61	5.36	6.40	3.47	
	MALE	5.31	4.00	4.68	2.13	
	ALL	6.58	4.55	5.69	3.06	

Display 16.13

Now the table has the right number of columns and rows, but it still doesn't work quite right. This is because there are extra spaces in each of the table cells.

If you delete them, the columns will line up properly.

Average years employment	Managerial	Professional	Technical	Sales
Non-Union FEMALE 7.36		6.19	5.83	4.81
MALE 6.56		5.29	4.18	3.48
ALL 6.93		5.71	4.96	4.27
Union FEMALE 7.61		5.36	6.40	3.47
MALE 5.31		4.00	4.68	2.13
ALL 6.58		4.55	5.69	3.06

Display 16.14

After you have removed the extra spaces, you can use the `Autoformat` option on the `Table` menu to choose and apply a table format, creating an attractive table like the one in Display 16.14.

Average years employment		Managerial	Professional	Technical	Sales
Non-Union	FEMALE	7.36	6.19	5.83	4.81
	MALE	6.56	5.29	4.18	3.48
	ALL	6.93	5.71	4.96	4.27
Union	FEMALE	7.61	5.36	6.40	3.47
	MALE	5.31	4.00	4.68	2.13
	ALL	6.58	4.55	5.69	3.06

This approach to importing a table works fine if you're working with a small number of simple tables, but if you are working with large numbers of tables, or if your tables are very complex, you may wish to try another approach.

One possibility is to automate more of the process. Chapter 13 in *Reporting From the Field: SAS Software Experts Present Real-World Report-Writing Applications* shows one method for doing this.

Another approach is outlined later in this chapter. You can output your table to an HTML file, and then import the HTML table into Word.

In Version 7, there is an experimental option to output your PROC TABULATE table to a Rich Text Format (.RTF) file that can be read directly into a word processor. This option will not be fully operational until Version 8. See Chapter 24 for more information.

Exporting Tables to HTML

As more and more information is delivered via the Internet, it's important to be able to produce reports not just on paper, but also in a format where they can be published on a Web page. SAS provides a number of tools to help you do this.

If you're going to put a lot of SAS output on the Web, you may want to investigate the SAS/IntrNet product. But if you just want to publish your PROC TABULATE tables, then you can download a free macro tool to do the trick. The HTML PROC TABULATE Formatter enables you to present any PROC TABULATE output as an HTML-formatted table.

To download the Formatter, go to http://www.sas.com, click on demos/downloads, and select HTML Formatting Tools (Version 1.2) from the list of alphabetical downloads. From this page, you will be able to download the Formatter, as well as the technical documentation to accompany it. Basically, the Formatter is a macro called %TAB2HTM that takes SAS output and creates an HTML file that you can publish on your Web site. Follow the instructions to install the Formatter, and then try out the following setup on your favorite PROC TABULATE table. This code produces a simple HTML table.

```
%TAB2HTM (CAPTURE=ON, RUNMODE=B);
   OPTIONS FORMCHAR='82838485868788898A8B8C'X;
   PROC TABULATE DATA=TEMP;
      CLASS OCCUP UNION GENDER;
      VAR EMPYEARS;
      TABLE UNION=' '*(GENDER=' ' ALL),
         OCCUP=' '*EMPYEARS=' '*MEAN=' '*F=12.2
         / BOX='Average years employment' ROW=FLOAT RTS=32;
   RUN;
%TAB2HTM(CAPTURE=OFF, RUNMODE=B, OPENMODE=REPLACE,
   HTMLFILE=C:\TEMP\TEST1.HTML, BRTITLE=PROC TABULATE as HTML);
```

This is a very basic setup. What it does is call %TAB2HTM once to start capturing the PROC TABULATE output, and then it calls %TAB2HTM again to end the capture. In the second call, a few parameters are used to control where the HTML file goes and what it looks like. Also, be sure to notice the strange FORMCHAR setting. This is a required setting — if you don't use this exact setting, %TAB2HTM will not work!

This example shows just a few of the many %TAB2HTM parameters. CAPTURE=ON and CAPTURE=OFF start and stop the capture of your SAS output. RUNMODE=B is used to indicate that this code will be run in batch mode.[3] OPENMODE=REPLACE indicates that the current output should overwrite any previous output in the file that is specified by HTMLFILE=' '. BRTITLE specifies a title to display in the browser window title bar. The resulting file, TEST1.HTML, is shown in Display 16.15. A Web browser was used to display the file.

There are many more %TAB2HTM parameters than the few presented here. A detailed explanation of each of the parameters is beyond the scope of this book. Consult the documentation that comes with the Formatter for more information.

---

[3] If you are running SAS Release 6.12, there is an interactive mode that you can use to create the HTML file and specify the %TAB2HTM properties that control table formatting.

Display 16.15

Netscape - [TABULATE as HTML]

File Edit View Go Bookmarks Options Directory Window Help

Location: file:///C|/TEMP/TEST1.HTML

Average years employment		Managerial	Professional	Technical	Sales
Non-Union	FEMALE	7.36	6.19	5.83	4.81
	MALE	6.56	5.29	4.18	3.48
	ALL	6.93	5.71	4.96	4.27
Union	FEMALE	7.61	5.36	6.40	3.47
	MALE	5.31	4.00	4.68	2.13
	ALL	6.58	4.55	5.69	3.06

Document: Done

If you are using SAS Version 7, you do not need to use %TAB2HTM to create HTML output. You can use the Output Delivery System instead. See Chapter 24 for an example of HTML output under Version 7.

Adding Hyperlinks to HTML Output

The great thing about HTML files is the way they can be linked to each other. By using hyperlinks, you can direct people from one page of your Web site to another, or even direct them to another Web site altogether.

While %TAB2HTM doesn't have a way to create these hyperlinks, there's no reason why you can't add them yourself. If you know a little about HTML code, it's easy to add a hyperlink to your table.

The following example shows how you can add a footnote to your table that directs the reader to the Web site that is the source of your data (in this case, the Census Bureau). All you have to know about HTML code is the following.

To create a hyperlink, you use two tags. The first tag is <A HREF='the URL goes here'>, and the second tag is </A>. These two tags mark the start and end of the text of the hyperlink. Whatever text you put between them is the text that will be displayed on your Web page. The URL that is listed in the HREF=' ' parameter is where the reader will go when they click on the hyperlink.

So to create an HTML tag for a hyperlink that is displayed as 'Source: Census Bureau' and will link the reader to the U.S. Census Bureau Web site, you use the following HTML code:

```
Source: <A HREF="http:\www.census.gov">Census Bureau</A>
```

To use this HTML code in your SAS program, add this text as a footnote to your table. You also need to make one change to the %TAB2HTM call. There is a parameter called ENCODE that determines what the macro does with angle brackets '<' and '>' like we've just used in our footnote. To get %TAB2HTM to leave the brackets alone and pass the footnote text directly to the HTML output file, you need to add ENCODE=N to the parameter list, as in the following code:

```
%TAB2HTM (CAPTURE=ON, RUNMODE=B);
   FOOTNOTE 'Source: <A HREF='http:\www.census.gov'>Census Bureau</A>';
   OPTIONS FORMCHAR='82838485868788898A8B8C'X;
   PROC TABULATE DATA=TEMP;
      CLASS OCCUP UNION GENDER;
      VAR EMPYEARS;
      TABLE UNION=' '*(GENDER=' ' ALL),
         OCCUP=' '*EMPYEARS=' '*MEAN=' '*F=12.2
         / BOX='Average years employment' ROW=FLOAT RTS=32;
   RUN;
%TAB2HTM(CAPTURE=OFF, RUNMODE=B, OPENMODE=REPLACE, ENCODE=N,
   HTMLFILE=C:\TEMP\TEST2.HTML, BRTITLE=PROC TABULATE as HTML);
```

The resulting table is shown in Display 16.16. Notice how the part of the footnote that was included between the angle brackets in the FOOTNOTE statement is now underlined. And if you click on this part of the footnote in a browser, you will be linked to www.census.gov.

This technique makes it easy to automate adding hyperlinks to your footnotes. You could also use this approach to put hyperlinks in titles, row headings, or column headings. This will work anywhere in the table where you can attach text labels.

Display 16.16

Netscape - [TABULATE as HTML]

File   Edit   View   Go   Bookmarks   Options   Directory   Window   Help

Location: C:\temp\TEST2.HTML

Average years employment		Managerial	Professional	Technical	Sales
Non-Union	FEMALE	7.36	6.19	5.83	4.81
	MALE	6.56	5.29	4.18	3.48
	ALL	6.93	5.71	4.96	4.27
Union	FEMALE	7.61	5.36	6.40	3.47
	MALE	5.31	4.00	4.68	2.13
	ALL	6.58	4.55	5.69	3.06

Source: Census Bureau

If this HTML stuff is confusing to you, don't worry about it. You don't have to understand how the HTML tags work to use this example. Just copy the code and try it out. This will make a lot more sense after you see it in action.

Converting PROC TABULATE Output to a Word Processor Table via HTML

There is another way to get your output into a word processor table beside importing it directly, as was shown in a previous example. These days, most word processors can read HTML files. Not only that, but they can turn HTML tables into word processor tables.

So, because we know how to create an HTML table from a PROC TABULATE table, it should be relatively easy to go from PROC TABULATE to a word processor table via an HTML table. This turns out to be the case, though there are a few quirks to formatting the table that will be shown in the following example.

The first thing you need to do is output your table to HTML. This example will use the basic HTML table we've created previously. To view the table, refer to Display 16.15.

After you have your HTML file, you need to open it in your word processor. The following screen images show what happens when you import a PROC TABULATE HTML table into Microsoft word.

Display 16.17

The first thing you will notice in Display 16.17 is that you have a problem with column widths.

When you first import the file, the columns are much too narrow.

To fix this problem, you need to reset the column widths. Put the cursor anywhere inside the table, and then from the menu bar click on `Table` and then `Select Table`. Now, on that same `Table` menu, select `Cell Height and Width`.

In the dialog box that comes up, select the column tab, and then reset all of the columns to a standard width (say 1 inch). You can revise this width later.

Display 16.18

In Display 16.18, you can now see all of the data in the table. Unfortunately, something appears to be wrong with the header. When we made all of the columns the same width, we ended up with the leftmost column header being one column wide instead of two columns.

To fix this, use the mouse to click and drag the column margins in the first row until everything lines up.

Average years employment		Managerial	Professional	Technical	Sales
Non-Union	FEMALE	7.36	6.19	5.83	4.81
	MALE	6.56	5.29	4.18	3.48
	ALL	6.93	5.71	4.96	4.27
Union	FEMALE	7.61	5.36	6.40	3.47
	MALE	5.31	4.00	4.68	2.13
	ALL	6.58	4.55	5.69	3.06

Display 16.19

Your table should now look pretty much like the HTML file you started with.

At this point, as you can see in Display 16.19, the table is looking pretty good. You could leave it like this. But if you want to clean it up further, turn to the next page.

Average years employment		Managerial	Professional	Technical	Sales
Non-Union	FEMALE	7.36	6.19	5.83	4.81
	MALE	6.56	5.29	4.18	3.48
	ALL	6.93	5.71	4.96	4.27
Union	FEMALE	7.61	5.36	6.40	3.47
	MALE	5.31	4.00	4.68	2.13
	ALL	6.58	4.55	5.69	3.06

Right now, the alignment of the text in the table headings and cells could use a little work. This is easy to do.

Unlike the example earlier in the chapter where the table ended up with a bunch of extra spaces in the labels, this time the text in each heading and cell is just the text, with no leading or trailing spaces.

Display 16.20

Average years employment		Managerial	Professional	Technical	Sales
Non-Union	FEMALE	7.36	6.19	5.83	4.81
	MALE	6.56	5.29	4.18	3.48
	ALL	6.93	5.71	4.96	4.27
Union	FEMALE	7.61	5.36	6.40	3.47
	MALE	5.31	4.00	4.68	2.13
	ALL	6.58	4.55	5.69	3.06

So to clean up the row headings, all we need to do is select these cells and then left-align them. To clean up the table cells, we can adjust the table cell spacing to put a little space between the right-aligned values and the cell borders. We can make the headers look sharper by choosing a black background and a white font to create a reverse effect. Finally, we can choose table borders that look less like HTML and more like a word processing document.

The resulting table, shown in Display 16.20, looks much more professional. And this is a simple example. There are many other ways you could dress up this table.

This chapter has now presented two ways to get PROC TABULATE output into a word processor. Which one should you use? The answer will depend on your word processor and your table. The best advice is to try it both ways. If you can get your table to work importing it directly from SAS into the word processor, then use that approach. However, if you end up doing a lot of reformatting of your output, you may want to try the HTML approach instead.

By the way, the HTML approach will also work for moving your table into a spreadsheet. Most spreadsheet applications can read HTML tables.

PART 5

Common Errors

CHAPTER 17

Troubleshooting Your Table

If you're having trouble getting a PROC TABULATE table to work, it's easy to waste a lot of time trying to see what went wrong. You can end up spending hours staring at the code trying to figure out the problem. The next time you run into difficulty, try using the following step-by-step approach to zero in on your problem.

1. **Are there any ERROR or WARNING messages in your log? Or are there any unusual NOTE messages in your log?**

   You're in luck. These are the easiest problems to fix. There are a few common error messages that come up all the time with PROC TABULATE. Refer to Chapter 18, "Error Messages and What They Mean," to see if your problem is listed. If so, you should be able to follow the example problem and solution and be able to fix your table. If this doesn't work for you, go on to step 2.

2. **Refer to the list of common problems in Chapter 19, "Incorrect Tables and How to Fix Them." Is your situation listed there?**

   Chapter 19 lists the most common problems that people run into when using PROC TABULATE. These are problems that do not generate an error message but do cause your table to be incorrect.

3. **Most problems can be solved by reading chapters 18 and 19. If you've got a more stubborn problem, turn the page to learn how to break your problem down step-by-step.**

   PROC TABULATE code can be confusing to debug because TABLE statements are so complex. The following pages show you how to break a table down into its components, so you can see what's going wrong.

4. Is your table fairly simple in design?

If your table is not too complicated, you may be able to get more help from SAS in building the table. You can use SAS/ASSIST to build a PROC TABULATE table without having to write any code. In SAS/ASSIST, you select your row variables, column variables, analysis variables, and statistics, and then SAS writes the code for you.

By using SAS/ASSIST and then comparing the code generated by SAS to the original code that you wrote, you may be able to figure out what you did wrong. Chapter 23, "Creating Tables Using SAS/ASSIST Software," explains in detail how to use SAS/ASSIST to generate your table.

5. If your table is not simple in design, the next step in debugging is to make the table simpler. Remove all of the special labels, formats, and options from the TABLE statement.

Sometimes the problem with a table is obvious if you look at the bare bones of the code. For example, the following table has a problem. As you can see in Output 17.1, every cell in the output table has the same value. This could be coincidence, but it's more likely that something is wrong.

```
PROC TABULATE DATA=TEMP;
   CLASS OWNHOME FULLTIME RACE AGE;
   VAR NUMKIDS;
   TABLE OWNHOME FULLTIME RACE,
      AGE*NUMKIDS=' '*MEAN*F=8.2
      / BOX='Number of Children' ROW=FLOAT RTS=32;
RUN;
```

The first step in finding the problem is to simplify this code, so we can see what's going on. We need to get rid of all of the extra stuff that isn't necessary to produce the basic table. This means getting rid of all of the TABLE statement options and removing all custom labels and formatting. What's left is the heart of the table.

```
PROC TABULATE DATA=TEMP;
   CLASS OWNHOME FULLTIME RACE AGE;
   VAR NUMKIDS;
   TABLE OWNHOME FULLTIME RACE,
      AGE*NUMKIDS*MEAN;
RUN;
```

Now try the running the table again. Maybe it was just one of the fancy formatting options that caused the problem. Looking at Output 17.2, you can see that the table is now a lot uglier and it still has the same problem. Every cell value is exactly the same as before.

If your table, like our example table, still has problems after applying these same simplifications, turn the page and go on to the next step.

Output 17.1

```
-----------------------------------------------------------------
|Number of Children               |            Age            |
|                                 |---------------------------|
|                                 | 20-39 | 40-59 | 60+       |
|                                 |-------+-------+-------    |
|                                 | MEAN  | MEAN  | MEAN      |
|---------------------------------+-------+-------+-------    |
|Home Ownership                   |       |       |           |
|---------------------------------|       |       |           |
|Owns Home                        | 3.00  | 3.00  | 3.00      |
|---------------------------------+-------+-------+-------    |
|Employment                       |       |       |           |
|---------------------------------|       |       |           |
|Full time                        | 3.00  | 3.00  | 3.00      |
|---------------------------------+-------+-------+-------    |
|Part time                        | 3.00  | 3.00  | 3.00      |
|---------------------------------+-------+-------+-------    |
|Race                             |       |       |           |
|---------------------------------|       |       |           |
|White                            | 3.00  | 3.00  | 3.00      |
|---------------------------------+-------+-------+-------    |
|Black                            | 3.00  | 3.00  | 3.00      |
|---------------------------------+-------+-------+-------    |
|Other                            | 3.00  | 3.00  | 3.00      |
-----------------------------------------------------------------
```

Output 17.2

```
-------------------------------------------------------------------
|                         |               Age                     |
|                         |---------------------------------------|
|                         | 20-39     | 40-59     | 60+           |
|                         |-----------+-----------+-----------    |
|                         | No. of    | No. of    | No. of        |
|                         | Children  | Children  | Children      |
|                         |-----------+-----------+-----------    |
|                         | MEAN      | MEAN      | MEAN          |
|-------------------------+-----------+-----------+-----------    |
|Home Ownership           |           |           |               |
|-------------------------|           |           |               |
|Owns Home                | 3.00      | 3.00      | 3.00          |
|-------------------------+-----------+-----------+-----------    |
|Employment               |           |           |               |
|-------------------------|           |           |               |
|Full time                | 3.00      | 3.00      | 3.00          |
|-------------------------+-----------+-----------+-----------    |
|Part time                | 3.00      | 3.00      | 3.00          |
|-------------------------+-----------+-----------+-----------    |
|Race                     |           |           |               |
|-------------------------|           |           |               |
|White                    | 3.00      | 3.00      | 3.00          |
|-------------------------+-----------+-----------+-----------    |
|Black                    | 3.00      | 3.00      | 3.00          |
|-------------------------+-----------+-----------+-----------    |
|Other                    | 3.00      | 3.00      | 3.00          |
-------------------------------------------------------------------
```

6. **If the table still doesn't work correctly, it's time to simplify again. Cut everything out of the PROC TABULATE code except the column dimension and the variables used in the column dimension.**

   PROC TABULATE problems can be caused by the something in the rows, something in the columns, or the interaction of the rows and columns. To find the problem, we need to look at each of these possibilities separately.

   We will first take a look at the column dimension. To do this, you take everything out of the TABLE statement except the column dimension. Then, remove everything from the CLASS and VAR statements that is not used in the TABLE statement. Our example code now looks like this:

   ```
   PROC TABULATE DATA=TEMP;
      CLASS AGE;
      VAR NUMKIDS;
      TABLE AGE*NUMKIDS*MEAN;
   RUN;
   ```

   The output from this simplified code is shown in Output 17.3. It appears that whatever was wrong with the table is now fixed, at least when there is only a column dimension.

   If your table is also working at this point, then go on to the next step. If not, try simplifying your column definition even more. If you have multiple CLASS variables, try running them one a time. If you have an analysis variable in the column dimension, try removing it and changing the statistic to an N.

   The goal is to keep simplifying until you've got the table working, even if only for one variable.

7. **After you have a part of the table working, start adding the rest of the TABLE definition back in, one variable at a time.**

   The goal is to figure out which addition will trigger the problem. After you know what causes the table to go wrong, you can examine the variable in more detail to figure out what's happening. For our example table, we know that the code works fine without the row dimension. So now we will take the part of the table that works and add in the part that doesn't work, one variable at a time.

   ```
   PROC TABULATE DATA=TEMP;
      CLASS AGE OWNHOME;
      VAR NUMKIDS;
      TABLE OWNHOME, AGE*NUMKIDS*MEAN;
   RUN;
   ```

   Looking at Output 17.4, you can see that our problem is back. And now we know where it originates. The only thing we changed in this code was the variable OWNHOME. So the problem appears to be with that variable.

   After checking the data by running a PROC FREQ, we discover that there was an error in calculating OWNHOME that caused it to have a missing value for most observations. The table looked odd because it was being run on very few observations. After fixing this data problem, the table was rerun, producing the correct output shown in Output 17.5.

   Your problem may not be spotted this quickly, but if you are patient about building your table step-by-step, you should be able to find the problem. If not, turn the page and try the next step.

Output 17.3

```
-----------------------------------------------------------
|                         Age                             |
|--------------------------------------------------------|
|   20-39      |    40-59     |     60+                   |
|------------- +------------- +-------------              |
|   No. of     |   No. of     |   No. of                  |
|  Children    |  Children    |  Children                 |
|------------- +------------- +-------------              |
|    MEAN      |    MEAN      |    MEAN                    |
|------------- +------------- +-------------              |
|         1.49 |         1.51 |         1.44              |
-----------------------------------------------------------
```

Output 17.4

```
----------------------------------------------------------------------------
|                          |                       Age                      |
|                          |-----------------------------------------------|
|                          |   20-39     |    40-59    |     60+           |
|                          |------------ +------------ +------------        |
|                          |   No. of    |   No. of    |   No. of          |
|                          |  Children   |  Children   |  Children         |
|                          |------------ +------------ +------------        |
|                          |    MEAN     |    MEAN     |    MEAN            |
|--------------------------+------------ +------------ +------------        |
|Home Ownership            |             |             |                   |
|--------------------------|             |             |                   |
|Owns Home                 |        3.00 |        3.00 |        3.00       |
----------------------------------------------------------------------------
```

Output 17.5

```
-----------------------------------------------------------
|Number of Children        |              Age             |
|                          |-----------------------------|
|                          | 20-39 | 40-59 |  60+        |
|                          |-------+-------+------        |
|                          | MEAN  | MEAN  | MEAN        |
|--------------------------+-------+-------+------        |
|Home Ownership            |       |       |             |
|--------------------------|       |       |             |
|Owns Home                 | 1.42  | 1.50  | 1.54        |
|--------------------------+-------+-------+------        |
|Rents Hom                 | 1.49  | 1.50  | 1.44        |
|--------------------------+-------+-------+------        |
|Employment                |       |       |             |
|--------------------------|       |       |             |
|Full time                 | 1.46  | 1.50  | 1.40        |
|--------------------------+-------+-------+------        |
|Part time                 | 1.65  | 1.43  | 1.83        |
|--------------------------+-------+-------+------        |
|Race                      |       |       |             |
|--------------------------|       |       |             |
|White                     | 1.50  | 1.51  | 1.46        |
|--------------------------+-------+-------+------        |
|Black                     | 1.64  | 1.40  | 1.50        |
|--------------------------+-------+-------+------        |
|Other                     | 1.25  | 1.49  | 1.47        |
-----------------------------------------------------------
```

8. **At this point, you are probably pretty frustrated. Now is the time to bring in expert assistance.**

Sometimes the solution to your problem is obvious, but you've been staring at your code so long you just can't see what's going on. Before you get hopelessly frustrated, try getting someone else to look at your code.

If there are other SAS programmers in your organization, this is the first resource to try. If you've got an expert to consult, that's great, but sometimes even a beginner can help you out. Often, your programming problems are very easy to spot — by someone else. Debugging your own code is always harder than debugging someone else's code.

If you don't have any other SAS programmers in your organization, now is the time to bring in SAS Technical Support. You can call SAS to talk directly with a Technical Support staff member, but with PROC TABULATE, it can be hard to explain over the phone what is going on. A better approach, if you have access to the Internet, is to send in your problem via e-mail.

Connect to www.sas.com and go to the Technical Support area, where you will find directions on how to submit your problem via e-mail. Type up a description of your problem, and then attach the following information:

- your PROC TABULATE code

- any relevant DATA step or PROC FORMAT code that applies to your table

- the log file produced by your code

- the output (if any) produced by your code.

Someone will examine your problem and contact you directly once they've sorted out the situation.

CHAPTER 18

Error Messages and What They Mean

It's not unusual to get ERROR messages in your SAS log the first time you run a new PROC TABULATE table. Tables can be complex, and it takes only one minor typographical mistake to create an ERROR message instead of the output you wanted.

PROC TABULATE has a few basic ERROR messages that you'll see again and again. This chapter explains each of them in detail, with examples of tables that can trigger each message, and tells you what to do to fix the problem.

"Type of name is unknown"

When you get a message saying that the "ERROR: The type of name (_____) is unknown," you are in luck. This is one of the easiest errors to fix. When you get this message, it means that you have a problem with one of the variables in the TABLE statement. PROC TABULATE even tells you which one is causing the problem. For example, the following code produces the error message in the log file shown in Output 18.1:

```
PROC TABULATE DATA=TEMP;
   CLASS EDUC;
   VAR INCOME;
   TABLE EDUC RACE, INCOME*MEAN;
RUN;
```

The error message says "The type of name (RACE) is unknown." This means that the TABLE statement is calling for a row variable called RACE, but this variable is not listed in either the CLASS statement or the VAR statement. This code is fixed by adding RACE to the CLASS statement. Because PROC TABULATE handles classification and analysis variables differently, every variable you use in a table has to be declared in a CLASS or VAR statement. Even if the variable is part of the data set, PROC TABULATE doesn't know what to do with it until you declare it as a classification or analysis variable.

This next example shows a slightly different variation in this error message. The following code produces the error messages shown in Output 18.2:

```
PROC TABULATE DATA=TEMP;
   CLASS EDUC RACE;
   VAR INCOME;
   TABLE EDUC RACE, HOURLY*MEAN;
RUN;
```

In this case, two error messages are triggered. The first message is the key to the problem. The table uses a variable called HOURLY, but this variable is not in either the CLASS or VAR statements. The second error message is related to the first. Because SAS can't figure out what to do with HOURLY, it also can't figure out how to calculate the MEAN. If you add HOURLY to the VAR statement, both problems will be fixed.

There is a third situation that can trigger the "type of name (_____) is unknown" message, and it is illustrated in the following code. This code produces the error message shown in Output 18.3.

```
PROC TABULATE DATA=TEMP;
   CLASS EDUC RACE;
   VAR INCOME;
   TABLE EDUD RACE, INCOME*MEAN;
RUN;
```

In this case, the error is caused by a typo. The variable EDUC is included in the CLASS statement, but it is misspelled as EDUD in the TABLE statement. If you fix the spelling, the problem is solved. In general, whenever you see the "type of name (_____) is unknown" message, all you have to do is compare the variables in your TABLE statement to the variables in your CLASS and VAR statements, and the problem will be obvious.

Output 18.1

```
11  PROC TABULATE DATA=TEMP;
12          CLASS EDUC;
13          VAR INCOME;
14          TABLE EDUC RACE,
15                  INCOME*MEAN;
16  RUN;

ERROR: The type of name (RACE) is unknown.
NOTE: The SAS System stopped processing this step because of errors.
```

Output 18.2

```
19  PROC TABULATE DATA=TEMP;
20          CLASS EDUC RACE;
21          VAR INCOME;
22          TABLE EDUC RACE,
23                  HOURLY*MEAN;
24  RUN;

ERROR: The type of name (HOURLY) is unknown.
ERROR: Statistic other than N was requested without analysis variable in the following nesting :
EDUC * HOURLY * MEAN.
NOTE: The SAS System stopped processing this step because of errors.
```

Output 18.3

```
26  PROC TABULATE DATA=TEMP;
27          CLASS EDUC RACE;
28          VAR INCOME;
29          TABLE EDUD RACE,
30                  INCOME*MEAN;
31  RUN;

ERROR: The type of name (EDUD) is unknown.
NOTE: The SAS System stopped processing this step because of errors.
```

 If you are using Version 7, be aware that the error messages may have slightly different wording than the error messages shown in this chapter. The meaning of the messages will remain unchanged.

"Statistic other than N was requested without analysis variable"

This message is also easy to figure out. Whenever you see this message, you know one of two things happened. Either you forgot to list your analysis variable in a VAR statement, or you've attached your statistic keyword to the wrong variable.

The following code is supposed to produce a table of mean age by education and marital status, as the title says. Unfortunately, what it produces is the error messages shown in Output 18.4.

```
PROC TABULATE DATA=TEMP;
    CLASS AGE EDUC MARITAL;
    TABLE EDUC MARITAL, AGE*MEAN;
    TITLE 'Mean Age by Education and Marital Status';
RUN;
```

You can see that this code produces two error messages. The first message tells you what has gone wrong: "ERROR: Statistic other than N was requested without analysis variable in the following nesting: EDUC * AGE * MEAN".

What happened is that PROC TABULATE tried to compute a MEAN of age by education, but the two variables were defined as CLASS variables, and PROC TABULATE can't compute a mean of a classification variable. If you think about it, this makes sense. How could you compute a mean *category* of age? What you want to compute is a mean *value* of age, and this requires that age be defined as an analysis variable in a VAR statement. If you move AGE from the CLASS statement to a VAR statement, this error message will go away.

The next example illustrates the second way you can trigger the "Statistic other than N was requested without analysis variable" message. In the following code, the analysis variable, AGE, is properly declared in a VAR statement. But as you can see from Output 18.5, something is wrong.

```
PROC TABULATE DATA=TEMP;
    CLASS EDUC MARITAL;
    VAR AGE;
    TABLE EDUC AGE, MARITAL*MEAN;
    TITLE 'Mean Age by Education and Marital Status';
RUN;
```

In this case, the problem is that the MEAN keyword is in the wrong place. PROC TABULATE can't figure out what to do for the part of the table where the row variable is EDUC, the column variable is MARITAL, and the statistic is MEAN. What does EDUC*MARITAL*MEAN mean? PROC TABULATE can't compute a mean marital status category for each value of education or a mean education category for each category of marital status. To fix this problem, the TABLE statement needs to be rearranged. There are a number of ways you could do this:

```
TABLE EDUC MARITAL, AGE*MEAN;
TABLE EDUC*AGE, MARITAL*MEAN;
TABLE EDUC*N AGE*MEAN, MARITAL;
```

(Which way you choose to fix this problem depends on what you wanted from the table in the first place.)

Output 18.4

```
15   PROC TABULATE DATA=TEMP;
16         CLASS AGE EDUC MARITAL;
17         TABLE EDUC MARITAL, AGE*MEAN;
18         TITLE 'Mean Age by Education and Marital Status';
19   RUN;

ERROR: Statistic other than N was requested without analysis variable in the following nesting :
EDUC * AGE * MEAN.
ERROR: Statistic other than N was requested without analysis variable in the following nesting :
MARITAL * AGE * MEAN.
NOTE: The SAS System stopped processing this step because of errors.
```

Output 18.5

```
60   PROC TABULATE DATA=TEMP;
61         CLASS EDUC MARITAL;
62         VAR AGE;
63         TABLE EDUC AGE, MARITAL*MEAN;
64         TITLE 'Mean Age by Education and Marital Status';
65   RUN;

ERROR: Statistic other than N was requested without analysis variable in the following nesting :
EDUC * MARITAL * MEAN.
NOTE: The SAS System stopped processing this step because of errors.
```

"Multiple statistics associated with a single table cell"

There are lots of times when you want to build a table that has more than one statistic. You may wish to show a mean and standard deviation or a minimum and maximum. It's also very common to add the N statistic to a table with other statistics in order to show the sample size.

But when you start using multiple statistics, you will occasionally run into the "ERROR: There are multiple statistics associated with a single table cell in the following nesting: _____" message. This is an easy problem to sort out. The error message states the problem pretty clearly.

For example, the following code produces the error message shown in Output 18.6:

```
PROC TABULATE DATA=TEMP;
   CLASS EDUC MARITAL;
   VAR AGE;
   TABLE EDUC*N MARITAL, AGE*MEAN;
RUN;
```

The error message points out that there are multiple statistics requested for a single table cell. The message even identifies the part of the table with the problem: it's the cells for the row EDUC*N and the column AGE*MEAN. PROC TABULATE doesn't know how to compute the MEAN of an N or the N of a MEAN.

This problem is very easy to spot. There are statistics in both the row and column dimensions of this table. To fix the problem, move the N to the column dimension or the MEAN to the row dimension. To avoid this problem in the future, be sure that all of your statistics are either in the page dimension, the row dimension, or the column dimension exclusively.

This sounds pretty easy, so this problem should never come up, right? Actually, you can also get this message even when all of the statistics are in one dimension. For example, look at the following code, which produces the log message shown in Output 18.7:

```
PROC TABULATE DATA=TEMP;
   CLASS EDUC MARITAL RACE;
   VAR AGE;
   TABLE (MARITAL*N RACE)*EDUC*MEAN, AGE;
RUN;
```

In this case, all of the statistics are in the row dimension. The problem is that there are too many statistics in the row dimension. If you remove the parentheses and expand out the row dimension, you can see what's happening. You end up with MARITAL*N*EDUC*MEAN as the first part of the column dimension, as is noted in the ERROR message. When you've got a lot of parentheses and asterisks in a single dimension you have to watch out for this. All you have to do to check for this problem is rewrite your code without the parentheses, and see if any part of the table has two statistics keywords in a single asterisked grouping.

To fix the problem, either the row dimension needs to be reorganized or one statistic needs to be dropped. Here are two ways this problem could be fixed:

```
TABLE (MARITAL RACE)*EDUC*MEAN, AGE;
TABLE MARITAL*N (MARITAL RACE)*EDUC*MEAN, AGE;
```

Output 18.6

```
16   PROC TABULATE DATA=TEMP;
17          CLASS EDUC MARITAL;
18          VAR AGE;
19          TABLE EDUC*N MARITAL, AGE*MEAN;
20   RUN;

ERROR: There are multiple statistics associated with a single table cell in the following nesting :
EDUC * N * AGE * MEAN.
NOTE: The SAS System stopped processing this step because of errors.
```

Output 18.7

```
11   PROC TABULATE DATA=TEMP;
12          CLASS EDUC MARITAL RACE;
13          VAR AGE;
14          TABLE (MARITAL*N RACE)*EDUC*MEAN, AGE;
15   RUN;

ERROR: There are multiple statistics associated with a single table cell in the following nesting :
MARITAL * N * EDUC * MEAN * AGE.
NOTE: The SAS System stopped processing this step because of errors.
```

"Multiple analysis variables associated with a single table cell"

Just as you often construct tables with more than one statistic, you may also want to construct tables with more than one analysis variable. You may want to show means for both age and income or for the sums of both number of employees and annual sales.

But when you start using multiple analysis variables, you will occasionally run into the "ERROR: There are multiple analysis variables associated with a single table cell in the following nesting: _____" message. This problem is very similar to the "multiple statistics" errors shown in the previous example. The following code shows how this problem occurs. It produces the error message shown in Output 18.8.

```
PROC TABULATE DATA=TEMP;
   CLASS EDUC MARITAL;
   VAR AGE INCOME;
   TABLE EDUC*INCOME MARITAL, AGE*MEAN;
RUN;
```

The error message points out that there are multiple analysis variables requested for a single table cell. The message even identifies the part of the table with the problem, it's the cells for the row EDUC*INCOME and the column AGE*MEAN. PROC TABULATE doesn't know how to compute the simultaneous mean of two analysis variables: AGE and INCOME.

This problem is very easy to spot. There are analysis variables in both the row and column dimensions of this table. To fix the problem, move INCOME to the column dimension or AGE to the row dimension. To avoid this problem in the future, be sure that all of your analysis variables are either in the page dimension, the row dimension, or the column dimension exclusively.

Actually, you can also get this message even when all of the analysis variables are in one dimension. For example, look at the following code, which produces the error message shown in Output 18.9:

```
PROC TABULATE DATA=TEMP;
   CLASS EDUC MARITAL RACE;
   VAR AGE INCOME;
   TABLE (MARITAL*INCOME RACE)*AGE*MEAN, EDUC;
RUN;
```

In this case, all of the analysis variables are in the row dimension. The problem is that there are too many analysis variables in the row dimension. If you remove the parentheses and expand out the row dimension, you can see what's happening. You end up with MARITAL*INCOME*AGE*MEAN as the first part of the row, as is noted in the ERROR message. When you've got a lot of parentheses and asterisks in a single dimension, you have to watch out for this. To check for this problem, rewrite your code without the parentheses, and see if any part of the table has two analysis variables in a single asterisked grouping.

To fix the problem, either the row dimension needs to be reorganized or one variable needs to be dropped. Here are two ways this problem could be fixed:

```
TABLE (MARITAL RACE)*(AGE INCOME)*MEAN, EDUC;
TABLE (MARITAL*INCOME RACE*AGE)*MEAN, EDUC;
```

Output 18.8

```
13    PROC TABULATE DATA=TEMP;
14          CLASS EDUC MARITAL;
15          VAR AGE INCOME;
16          TABLE EDUC*INCOME MARITAL, AGE*MEAN;
17    RUN;
```

ERROR: There are multiple analysis variables associated with a single table cell in the following
nesting : EDUC * INCOME * AGE * MEAN.
NOTE: The SAS System stopped processing this step because of errors.

Output 18.9

```
58    PROC TABULATE DATA=TEMP;
59          CLASS EDUC MARITAL RACE;
60          VAR AGE INCOME;
61          TABLE (MARITAL*INCOME RACE)*AGE*MEAN, EDUC;
62    RUN;
```

ERROR: There are multiple analysis variables associated with a single table cell in the following
nesting : MARITAL * INCOME * AGE * MEAN * EDUC.
NOTE: The SAS System stopped processing this step because of errors.

"Class variable is missing on every observation"

The previous examples show problems that come up because of faulty PROC TABULATE code. This message usually refers to a problem with the data set itself. When you see the message "WARNING: A class, frequency, or weight variable is missing on every observation", you know that you have a DATA step problem.

This message is only a WARNING, but it might as well be an ERROR, because PROC TABULATE cannot build a table when there is a missing CLASS value for every observation. To see how this problem occurs, take a look at the following code, which produces the error message shown in Output 18.10:

```
PROC TABULATE DATA=TEMP;
   CLASS APPR MTRAIN UNION OCCUP;
   TABLE OCCUP=' ',
   (APPR MTRAIN)*UNION=' '*N=' '*F=10.
   / ROW=FLOAT BOX='Programs Offered' RTS=20;
RUN;
```

Unfortunately, this error message doesn't give you much help. All you know is that one or more of the CLASS variables has so much missing data that there are no observations where each of the CLASS variables has non-missing data.

You could look at the above code for hours and never figure out what's going wrong. To see what's causing the problem, you need to rerun the code with the MISSING option turned on. This will let you see the pattern of missing data so that you can fix the problem in the data.

Output 18.11 shows what happens when you add the MISSING option to the PROC TABULATE statement. Now you can see the problem clearly. The variable APPR is missing for all records where UNION='Union', and the variable MTRAIN is missing for all records where UNION='Non-Union'. So every record in the data set has missing data for either APPR or MTRAIN. It turns out that in this data set, only union employers offer apprenticeships, and only nonunion employers offer management-training programs.

To fix the problem, we can modify the data set to recode the missing data and then rerun the table without the MISSING option.

```
DATA FIXED;
   SET TEMP;
   IF UNION=1 THEN MTRAIN=0;
   ELSE IF UNION=0 THEN APPR=0;
RUN;
PROC TABULATE DATA=FIXED;
   CLASS APPR MTRAIN UNION OCCUP;
   TABLE OCCUP=' ',
   (APPR MTRAIN)*UNION=' '*N=' '*F=10.
   / ROW=FLOAT BOX='Programs Offered' RTS=20;
RUN;
```

You can see in Output 18.12 that the problem is now solved (though the table does not make much sense). Sometimes the best solution to this problem is to redesign the table instead of trying to fix the data.

Output 18.10

```
10   PROC TABULATE DATA=TEMP;
11          CLASS APPR MTRAIN UNION OCCUP;
12          TABLE OCCUP=' ',
13                 (APPR MTRAIN)*UNION=' '*N=' '*F=10.
14                 / ROW=FLOAT BOX='Programs Offered' RTS=20;
15   RUN;
```

WARNING: A class, frequency, or weight variable is missing on every observation.

Output 18.11

Programs Offered	Apprentice			Management Training		
	.	No	Yes	.	No	Yes
	Non-Union	Union	Union	Union	Non-Union	Non-Union
Managerial	209	10	10	20	102	107
Professional	373	49	44	93	201	172
Technical	57	8	9	17	32	25
Sales	102	5	8	13	48	54

Output 18.12

Programs Offered	Apprentice			Management Training		
	No	Yes		No	Yes	
	Non-Union	Union	Union	Non-Union	Union	Non-Union
Managerial	209	10	10	102	20	107
Professional	373	49	44	201	93	172
Technical	57	8	9	32	17	25
Sales	102	5	8	48	13	54

"PCTN base is not in table",
"PCTN crossing has no denominator"

If you're building tables with percentages, you're going to run into these messages sooner or later. These two messages always come together. They are caused by the same problem: your denominator definition does not match your table design. Whatever you put between the brackets of the PCTN<> or PCTSUM<> does not match up to the column, row, or cell that it is being applied to.

There are infinite ways you can trigger this message. This example illustrates one of the more common causes and its solution.

```
PROC TABULATE DATA=TEMP;
   CLASS MARITAL EDUC;
   TABLE MARITAL ALL, EDUC*PCTN<MARITAL>;
RUN;
```

This code triggers the error messages shown in Output 18.13. What's causing the problem is the variable ALL. This variable is used in the row dimension but it is not in the denominator definition. When PROC TABULATE tries to calculate the column percentage for each marital status, the denominator definition makes sense. PROC TABULATE can compute the percentage of the observations that are in each marital status. However, when PROC TABULATE comes to the final row of the table, the row for ALL, there is no denominator to use. To fix the problem, all you have to do is use the entire row definition as a denominator definition. The new TABLE statement is as follows:

```
TABLE MARITAL ALL, EDUC*PCTN<MARITAL ALL>;
```

Now the table is produced correctly, as shown in Output 18.14. In this simple table, it is fairly easy to spot and fix the problem. When you have a more complex table, it gets a lot harder.

```
TABLE MARITAL=" " RACE=" "*(GENDER="GENDER" ALL),
      EDUC*F=8.1*PCTN<RACE*GENDER>;
```

This sample TABLE statement, which generates the same error messages, illustrates a strategy for solving more complex denominator problems. The first step in solving the problem is to simplify your TABLE statement. Remove all formats, statistics (including PCTN), and labels. Get rid of any parentheses by multiplying them out. Now you should have a TABLE statement made up of variable names, asterisks, and commas.

```
TABLE MARITAL RACE*GENDER RACE*ALL, EDUC;
```

Next, figure out the list of possible denominators. One denominator that always works is if you use the entire row dimension, <MARITAL RACE*GENDER RACE*ALL>. This gives you column percentages. Another denominator that always works is the entire column dimension <EDUC>. This gives you row percentages. If you want cell percentages, just get rid of the denominator definition altogether.

You can also use any denominator that has one item from each of the components of the row or column definition (components are the items in the definition that are separated by spaces). In this case, there are no other possible denominators based on the column definition because there is only one variable (EDUC). But the row dimension has several possibilities: <MARITAL RACE>, <MARITAL GENDER ALL>, <MARITAL RACE ALL>, and <MARITAL GENDER RACE>. You can't use MARITAL, RACE, or GENDER alone as the denominator because none of these variables is used in all three components of the row definition.

Output 18.13

```
13    PROC TABULATE DATA=TEMP;
14         CLASS MARITAL EDUC;
15         TABLE MARITAL ALL, EDUC*PCTN<MARITAL>;
16    RUN;

ERROR: PCTN base is not in table.
ERROR: A PCTN crossing has no denominator.
```

Output 18.14

```
-----------------------------------------------------------------------
|                                |            Education                |
|                                |-------------------------------------|
|                                |   <HS     |    HS     |  College    |
|                                |-----------+-----------+-----------  |
|                                |   PCTN    |   PCTN    |   PCTN      |
|--------------------------------+-----------+-----------+-----------  |
|Marital Status                  |           |           |             |
|--------------------------------|           |           |             |
|Married                         |    67.10|      68.35|      67.61 |
|--------------------------------+-----------+-----------+----------- |
|Widowed                         |     3.26|       2.53|       1.64 |
|--------------------------------+-----------+-----------+----------- |
|Divorced/Sep.                   |    14.33|      14.81|      10.96 |
|--------------------------------+-----------+-----------+----------- |
|Never Married                   |    15.31|      14.31|      19.79 |
|--------------------------------+-----------+-----------+----------- |
|ALL                             |   100.00|     100.00|     100.00 |
-----------------------------------------------------------------------
```

For more information on complex denominators, refer to Chapter 8, "Percentages with Complex Denominators." In particular, check out the chart of denominator definitions in the example on "Picking the Right Denominator."

CHAPTER 19

Incorrect Tables and How to Fix Them

The previous chapter tackled problems that are easy to solve. When PROC TABULATE gives you an error message, it's usually not too hard to figure out the problem.

But sometimes you don't get an error message when something's wrong. You run your code, there's no indication of trouble in the log, but when you look at the output table, it's just plain wrong.

There are a number of common problems that can create incorrect output without generating an error message. This chapter addresses some of the more common problems and their solutions.

Number of Observations Seems too Small

Your code runs fine with no error messages, the table has the rows and columns you wanted, the values are formatted correctly, but the table is still wrong. What do you do when your table shows a number of observations that is much smaller than what you expected?

Most of the time, when you produce a table, you know how many observations to expect. In fact, one of the best ways to double-check your work is to include the number of observations in your table, at least for the first draft. After you've reviewed the table and the number of observations looks okay, you can always choose to delete the N statistic from the final report. In the following example, the N statistic is used to check the observation counts for a table showing mean hourly wages:

```
PROC TABULATE DATA=TEMP;
   CLASS UNION OCCUP;
   VAR HOURLY;
   TABLE OCCUP=' ' ALL='Total',
      UNION=' '*HOURLY=' '*(N*F=8. MEAN*F=DOLLAR10.2)
      / BOX='Hourly Wages' RTS=25;
RUN;
```

This data set happens to have over 4,000 observations, so we'd expect to see roughly that many observations in the output table. But if you look at Output 19.1, you can see that this table contains only about 450 observations.

To figure out what's causing the problem, there are two things you should try. They are shown separately here for clarity, but normally you should try both at once.

The first thing to do is look for missing values for the analysis variable HOURLY. To do this, add the NMISS statistic to your TABLE statement.

```
TABLE OCCUP=' ' ALL='Total',
   UNION=' '*HOURLY=' '*(N*F=8. NMISS*F=8. MEAN*F=DOLLAR10.2)
   / BOX='Hourly Wages' RTS=25;
```

Now the table, shown in Output 19.2, has a column showing how many observations did not have a value for HOURLY. As you can see, this accounts for some of the missing data, but we still have only about 550 observations. We're missing over 3,000 observations.

The other thing we need to look for is missing data for the CLASS variables. To do this, use the MISSING option in the PROC TABULATE statement. This causes missing values to be listed as separate categories in the row and column headings.

```
PROC TABULATE DATA=TEMP MISSING;
```

The table in Output 19.3 shows that this is our culprit. There are over 3,200 observations with no data for the variable UNION. To fix this table, we need to decide what to do with these observations. It may be possible to recode these observations, or they could be displayed with an appropriate label. For more information on how to handle missing data, see Chapter 9, "Handling Missing Data."

Output 19.1

Hourly Wages	Non-Union		Union	
	N	MEAN	N	MEAN
Managerial	105	$18.90	9	$15.30
Professional	184	$16.69	46	$17.27
Technical	31	$13.23	9	$17.22
Sales	54	$18.25	8	$14.75
Total	374	$17.25	72	$16.74

Output 19.2

Hourly Wages	Non-Union			Union		
	N	NMISS	MEAN	N	NMISS	MEAN
Managerial	105	26	$18.90	9	2	$15.30
Professional	184	53	$16.69	46	10	$17.27
Technical	31	5	$13.23	9	1	$17.22
Sales	54	13	$18.25	8	2	$14.75
Total	374	97	$17.25	72	15	$16.74

Output 19.3

Hourly Wages	.		Non-Union		Union	
	N	MEAN	N	MEAN	N	MEAN
Managerial	944	$18.17	105	$18.90	9	$15.30
Professional	1506	$16.44	184	$16.69	46	$17.27
Technical	311	$12.18	31	$13.23	9	$17.22
Sales	508	$16.30	54	$18.25	8	$14.75
Total	3269	$16.51	374	$17.25	72	$16.74

Some of the Values of the CLASS Variable Are Not Shown

Most of the time, when you create a table, you expect to see certain rows and columns in the result. When a variable has three categories, say 'red,' 'white,' and 'blue,' then you expect a table to show rows and columns for all three colors. If you get 'just red' and 'white,' then something is wrong.

Take the case of the following table. The row variable is GRADE. According to the variable's format, the values for this variable are A, B, C, D, and F.

```
PROC TABULATE DATA=TEMP;
   CLASS GRADE RACE;
   TABLE GRADE, RACE=' '*N*F=8. / RTS=10;
RUN;
```

But if you look at Output 19.4, you can see that the table has values for A, B, C, and F only. There are no records with a grade of D. To figure out what's going on, you can run a simple PROC FREQ on GRADE.

```
PROC FREQ DATA=TEMP;
   TABLES GRADE;
RUN;
```

As you can see in Output 19.5, the problem with our table is caused by the variable GRADE, not any mistake in the PROC TABULATE code. There are no D grades in the data set, so there's nothing for PROC TABULATE to display. This may be a sign of data problems, so at this point you should check the data set carefully.

If the data is correct, and there are no D grades in this population, you have two choices for how to proceed. You could leave the table alone and add a footnote explaining that there are no D grades. Or, you could use code like the following to force a row for the grade D into the table:

```
DATA TEMPFIX;
   GRADE='D'; RACE=1; OUTPUT;
   GRADE='D'; RACE=2; OUTPUT;
   GRADE='D'; RACE=3; OUTPUT;
RUN;
DATA TEMP2;
   SET TEMPFIX (IN=F1) TEMP;
   IF F1 THEN COUNTER=.; ELSE COUNTER=1;
RUN;
PROC TABULATE DATA=TEMP2;
   CLASS GRADE RACE;
   VAR COUNTER;
   TABLE GRADE, RACE=' '*COUNTER=' '*N*F=8. / RTS=10;
RUN;
```

This solution works, as shown in Output 19.6, but it's an ugly solution. There are other possible solutions that are shown in Chapter 9, "Handling Missing Data," and Chapter 15, "PROC TABULATE Tricks: How to Cheat to Create Complex Tables."

Output 19.4

```
----------------------------------------
|          | White  | Black  | Other  |
|          |--------+--------+--------|
|          |   N    |   N    |   N    |
|--------+--------+--------+--------|
|Grade     |        |        |        |
|--------|        |        |        |
|A        |    452 |     54 |     60 |
|--------+--------+--------+--------|
|B        |   1111 |     96 |    143 |
|--------+--------+--------+--------|
|C        |    995 |    113 |    130 |
|--------+--------+--------+--------|
|F        |    846 |     74 |    110 |
----------------------------------------
```

Output 19.5

GRADE	Frequency	Percent	Cumulative Frequency	Cumulative Percent
A	566	13.5	566	13.5
B	1350	32.3	1916	45.8
C	1238	29.6	3154	75.4
F	1030	24.6	4184	100.0

Output 19.6

```
----------------------------------------
|          | White  | Black  | Other  |
|          |--------+--------+--------|
|          |   N    |   N    |   N    |
|--------+--------+--------+--------|
|Grade     |        |        |        |
|--------|        |        |        |
|A        |    452 |     54 |     60 |
|--------+--------+--------+--------|
|B        |   1111 |     96 |    143 |
|--------+--------+--------+--------|
|C        |    995 |    113 |    130 |
|--------+--------+--------+--------|
|D        |      0 |      0 |      0 |
|--------+--------+--------+--------|
|F        |    846 |     74 |    110 |
----------------------------------------
```

 If you are running SAS Version 7, there is a better solution for this problem. See the examples on the PRELOADFMT, CLASSDATA, and EXCLUSIVE options in Chapter 24, "Version 7 Enhancements."

A Format Is Not Being Applied to a Variable

Unlike most SAS procedures, where there is only one way to apply formats to your output, PROC TABULATE has several ways to apply formats. First, you can use the FORMAT= option in the PROC TABULATE statement. Second, you can use the FORMAT= (or F=) option in the TABLE statement. And third, you can use a FORMAT statement.

Unfortunately, these three methods are not interchangeable. This can lead to a lot of confusion when your table does not turn out as expected. For example, look at the following two tables. One table has a format for the analysis variable INCOME. The second table has formats for the two class variables, OCCUP and RACE.

```
PROC TABULATE DATA=TEMP;
   CLASS EDUC AGE;
   VAR INCOME;
   TABLE EDUC, AGE*INCOME=' '*MEAN=' ' / BOX='Mean Income';
   FORMAT INCOME DOLLAR8.;
RUN;
PROC TABULATE DATA=TEMP;
   CLASS OCCUP RACE;
   TABLE OCCUP*F=OCCUP., RACE*F=NEWRACE.*N;
RUN;
```

But both of these tables are set up with the wrong type of format assignment. The result is that none of the formats are applied to the tables, shown in Output 19.7. In the first output table, you can see that the income variable is formatted as 12.2, not DOLLAR8. And in the second table, neither the OCCUP. or NEWRACE. formats have been applied.

The reason for this is that the FORMAT statement only applies to CLASS variable values, so it will not have any effect on the analysis variable INCOME. And the FORMAT= option or the F= format modifier only affects analysis variables, so they can not be used to format the classification variables OCCUP and RACE.

To fix the problem, we need to reverse the approach used in the two tables. INCOME is formatted using the F= modifier, and OCCUP and RACE are formatted with a FORMAT statement.

```
PROC TABULATE DATA=TEMP;
   CLASS EDUC AGE;
   VAR INCOME;
   TABLE EDUC, AGE*INCOME=' '*MEAN=' '*F=DOLLAR8.
      / BOX='Mean Income';
RUN;
PROC TABULATE DATA=TEMP;
   CLASS OCCUP RACE;
   TABLE OCCUP, RACE*N*F=8.;
   FORMAT OCCUP OCCUP. RACE NEWRACE.;
RUN;
```

The corrected tables are shown in Output 19.8. Now INCOME is shown with the DOLLAR8. format, and the classification variables OCCUP and RACE have their appropriate labels.

Output 19.7

Mean Income	Age		
	20-39	40-59	60+
Education			
<HS	16569.05	18652.46	23433.17
HS	23442.46	25038.63	25331.50
College	32463.98	41767.57	42933.07

	Race				
	1	2	3	4	5
	N	N	N	N	N
Occupation					
1	949	114	44	47	32
2	1606	143	79	70	56

Output 19.8

Mean Income	Age		
	20-39	40-59	60+
Education			
<HS	$16,569	$18,652	$23,433
HS	$23,442	$25,039	$25,332
College	$32,464	$41,768	$42,933

	Race		
	White	Black	Other
	N	N	N
Occupation			
Managerial	949	114	123
Professional	1606	143	205

Blank Label Is Not Being Applied to a Variable

Just as there is more than one way to apply a format in PROC TABULATE, there is also more than one way to apply a label. You can label variables in the TABLE statement, using the VARNAME=' ' syntax, or you can label variables using a LABEL statement. Similarly, you can label statistics in the TABLE statement or with a KEYLABEL statement.

Generally, these two approaches are interchangeable, but there is one exception. Take a look at the following example, which uses LABEL and KEYLABEL statements to apply labels to the variables in the TABLE statement:

```
PROC TABULATE DATA=TEMP F=13.1;
   CLASS MARITAL AGE;
   VAR NUMKIDS;
   TABLE AGE, NUMKIDS*MARITAL*MEAN;
   LABEL AGE='Age Group'
         NUMKIDS='Number of Children'
         MARITAL=' ';
   KEYLABEL MEAN=' ';
RUN;
```

When you look at the resulting table in Output 19.9, you can see that one of the labels did not get applied. The blank label was not applied to the variable MARITAL. This is the exception. If you want to apply a blank label to a variable name, you have to do so in the TABLE statement.

Notice that this rule does not apply to statistic labels. The blank label assigned to MEAN in the KEYLABEL statement is applied in the output table. You get the same result whether MEAN=' ' is in the TABLE statement or the KEYLABEL statement.

To fix the table, move the blank label for MARITAL into the TABLE statement, as in the following code:

```
PROC TABULATE DATA=TEMP F=13.1;
   CLASS MARITAL AGE;
   VAR NUMKIDS;
   TABLE AGE, NUMKIDS*MARITAL=' '*MEAN;
   LABEL AGE='Age Group'
         NUMKIDS='Number of Children';
   KEYLABEL MEAN=' ';
RUN;
```

The correct output is shown in Output 19.10. If you find it confusing to keep track of which labels can be applied where, the safest thing to do is just apply all labels in the TABLE statement, as in the following example. This will produce a table identical to Output 19.10.

```
TABLE AGE='Age Group',
   NUMKIDS='Number of Children'*MARITAL=' '*MEAN=' ';
```

Output 19.9

	Number of Children			
	MARITAL			
	Married	Widowed	Divorced/Sep.	Never Married
Age Group				
20-39	1.5	2.0	1.6	1.5
40-59	1.5	1.7	1.6	1.5
60+	1.4	1.7	1.5	1.5

Output 19.10

	Number of Children			
	Married	Widowed	Divorced/Sep.	Never Married
Age Group				
20-39	1.5	2.0	1.6	1.5
40-59	1.5	1.7	1.6	1.5
60+	1.4	1.7	1.5	1.5

A Statistic's Label Is Blank, but a Space Is Left In the Row Label

There's more to specifying a blank label than just putting VARNAME=' ' in the TABLE statement.

For example, the following code correctly specifies blank labels for both the variable HOURLY and the statistic MEAN.

```
PROC TABULATE DATA=TEMP;
   CLASS AGE RACE;
   VAR HOURLY;
   TABLE AGE*HOURLY=' '*MEAN=' ', RACE=' '
      / BOX='Mean Hourly Wage';
RUN;
```

But if you look at the table in Output 19.11, you can see that while the two labels are blank, the two boxes created to hold them were left in the row headings. This is because we forgot one step in removing labels.

The first thing you have to do is make them blank by assigning a VARNAME=' ' label. The second step is to remove the extra row headings from the grid by specifying the TABLE statement option ROW=FLOAT. The following code fixes the problem:

```
PROC TABULATE DATA=TEMP;
   CLASS AGE RACE;
   VAR HOURLY;
   TABLE AGE*HOURLY=' '*MEAN=' ', RACE=' '
      / BOX='Mean Hourly Wage' ROW=FLOAT;
RUN;
```

Now the table, shown in Output 19.12, no longer has those extra boxes. This problem could also have been solved by moving the statistics from the row dimension to the column dimension. PROC TABULATE will remove the boxes from the grid automatically for any column headings with blank labels. You should always use one of these techniques when you have blank labels for row variables or statistics.

Another thing you may want to do in this situation is adjust the row title space with the RTS option, as in the following code.

```
PROC TABULATE DATA=TEMP;
   CLASS AGE RACE;
   VAR HOURLY;
   TABLE AGE*HOURLY=' '*MEAN=' ', RACE=' '
      / BOX='Mean Hourly Wage' ROW=FLOAT RTS=20;
RUN;
```

Narrowing the row headings makes the table in Output 19.13 much easier to read. It also saves space, which could come in handy with a wider table.

Output 19.11

Mean Hourly Wage			White	Black	Other
Age					
20-39			15.45	15.31	16.65
40-59			17.16	18.53	16.99
60+			16.63	19.24	18.85

Output 19.12

Mean Hourly Wage	White	Black	Other
Age			
20-39	15.45	15.31	16.65
40-59	17.16	18.53	16.99
60+	16.63	19.24	18.85

Output 19.13

Mean Hourly Wage	White	Black	Other
Age			
20-39	15.45	15.31	16.65
40-59	17.16	18.53	16.99
60+	16.63	19.24	18.85

Statistics Labels Do Not Appear Next to the Correct Variable

You probably have your desired table pictured perfectly in your mind, but getting it to look like that in your SAS output can sometimes be a challenge. Unfortunately, PROC TABULATE cannot read your mind.

A common problem is that you get the basic structure of the table correct, and you get the correct values in the table cells, but the row and column headings are not as you would like. For example, take a look at the following code:

```
PROC TABULATE DATA=TEMP F=DOLLAR8.;
   CLASS AGE NUMKIDS;
   VAR INCOME;
   TABLE INCOME*AGE, NUMKIDS*MEAN;
RUN;
```

There's nothing wrong with the table shown in Output 19.14, but it is hard to read. It looks like it is displaying the mean number of children because the MEAN label is below the number of children headings in the column dimension. Actually, MEAN applies to the analysis variable INCOME.

While you may know that this table shows mean incomes, and PROC TABULATE has figured out that you wanted mean incomes, someone reading your table may get confused. The table would be much more understandable if MEAN was moved from the column headings to the row headings.

```
TABLE INCOME*AGE*MEAN, NUMKIDS;
```

Again, there's nothing wrong with the output produced by this TABLE statement. The table is shown in Output 19.15. But the table is still fairly hard to read. It looks like it is displaying mean ages, because the MEAN label is next to AGE in the row headings.

Moving the statistic label to the row heading was not enough to fix our problem. We have to also put in the right part of the row heading. We want it to end up next to the variable INCOME. The following TABLE statement should do the trick:

```
TABLE INCOME*MEAN*AGE, NUMKIDS;
```

Now the table, shown in Output 19.16, is much clearer. You can easily tell that it is a table of mean incomes. Notice that the numbers shown in the table have not changed from example to example. This has been the same table all along. The art to building good PROC TABULATE tables is in figuring out how to order the row and column headings and where to place the statistics so that the table is not just correct but is also easy to read.

There is one more step we could take with this example table. The three-level row headings are still a little hard to read, and they're not attractive. One option is to take the labels out of the rows and instead put a summary label in the table BOX, as in the following code:

```
TABLE INCOME=' '*MEAN=' '*AGE=' ', NUMKIDS
   / BOX='Mean Income by Age Group' ROW=FLOAT RTS=12;
```

The resulting table is shown in Output 19.17. It is both correct and easy to read.

Output 19.14

		No. of Children			
		0	1	2	3
		MEAN	MEAN	MEAN	MEAN
Income	Age				
	20-39	$29,063	$27,779	$27,723	$30,721
	40-59	$35,116	$33,320	$35,757	$33,995
	60+	$31,324	$34,829	$37,947	$32,734

Output 19.15

			No. of Children			
			0	1	2	3
Income	Age					
	20-39	MEAN	$29,063	$27,779	$27,723	$30,721
	40-59	MEAN	$35,116	$33,320	$35,757	$33,995
	60+	MEAN	$31,324	$34,829	$37,947	$32,734

Output 19.16

			No. of Children			
			0	1	2	3
Income	MEAN	Age				
		20-39	$29,063	$27,779	$27,723	$30,721
		40-59	$35,116	$33,320	$35,757	$33,995
		60+	$31,324	$34,829	$37,947	$32,734

Output 19.17

Mean Income by Age Group	No. of Children			
	0	1	2	3
20-39	$29,063	$27,779	$27,723	$30,721
40-59	$35,116	$33,320	$35,757	$33,995
60+	$31,324	$34,829	$37,947	$32,734

Statistics Are Grouped Incorrectly

The reason we build tables is to use them to ask questions of our data. Sometimes you build a perfectly good table, but because of the way it is organized, it's hard to figure out what's going on in the data.

This is the case with the following example. We have some data on tax rates, age, and education levels. What we want to know is do tax rates vary by age or education? Also, we want to know if there is more variation in tax rates in any particular age or education group. The statistics we need to answer these questions are MEAN and STD.

The following code produces a table with means and standard deviations for RATE by age and education level. The code is set up correctly, and there is nothing inherently wrong with the output.

```
PROC TABULATE DATA=TEMP;
   CLASS AGE EDUC;
   VAR RATE;
   TABLE AGE*RATE=' '*(MEAN STD), EDUC
      / BOX='Tax Rates' ROW=FLOAT;
RUN;
```

If you look at the resulting table in Output 19.18, you can indeed see the various means and standard deviations, but it's hard to compare them across combinations of AGE and EDUC. It is relatively easy to see differences in tax rate by education for each age group because you can compare the means across the rows and the standard deviations across rows.

But to look at it the other way, looking at differences by age within each education level, it is hard to read. As you compare down the column, the means are interspersed with standard deviations. If we want to look at the table this way, it would be better to reorganize the table with the statistics in the column dimension.

```
TABLE AGE*RATE=' ', EDUC*(MEAN STD)
   / BOX='Tax Rates' ROW=FLOAT RTS=10;
```

Now, in Output 19.19, it is easy to compare the means across age groups for each level of education. Which way you build this table will depend on how you want to look at things. There's a third way to build this table. If you want to look at the means by education and age group simultaneously, use the following code:

```
TABLE AGE*RATE=' ', (MEAN STD)*EDUC
   / BOX='Tax Rates' ROW=FLOAT RTS=10;
```

In Output 19.20, all of the means are grouped together, so you can look at the combinations of age and education any way you want. The catch with this approach is that it's hard to look at how the means and standard deviations work together.

Ultimately, you have to decide which way to build your table. The choice you make will depend on the questions you are trying to ask of your data. No matter what you're trying to find out, there's usually some way to reconfigure your table to make the answer clearer.

Output 19.18

Tax Rates		Education		
		<HS	HS	College
Age				
20-39	MEAN	0.32	0.33	0.34
	STD	0.06	0.05	0.05
40-59	MEAN	0.33	0.33	0.34
	STD	0.05	0.05	0.05
60+	MEAN	0.34	0.33	0.34
	STD	0.05	0.05	0.05

Output 19.19

Tax Rates	Education					
	<HS		HS		College	
	MEAN	STD	MEAN	STD	MEAN	STD
Age						
20-39	0.32	0.06	0.33	0.05	0.34	0.05
40-59	0.33	0.05	0.33	0.05	0.34	0.05
60+	0.34	0.05	0.33	0.05	0.34	0.05

Output 19.20

Tax Rates	MEAN			STD		
	Education			Education		
	<HS	HS	College	<HS	HS	College
Age						
20-39	0.32	0.33	0.34	0.06	0.05	0.05
40-59	0.33	0.33	0.34	0.05	0.05	0.05
60+	0.34	0.33	0.34	0.05	0.05	0.05

Something Is Wrong with the Column Headings

This problem does not come up very often, but it's quite a surprise when it does. You get a perfectly correct table with part of the grid missing.

The problem is caused by tables like the following:

```
PROC TABULATE DATA=TEMP F=8.;
   CLASS RACE EDUC AGE;
   VAR INCOME;
   TABLE AGE, ALL=' '*N=' '
      (RACE=' ' EDUC)*INCOME*MEAN=' '
      / RTS=10;
RUN;
```

As you can see from Output 19.21, there are lines missing from the table grid. This is caused by the definition of the table's first column. The column is supposed to show the number of observations for each age group. But you will notice in the code that the column definition ALL=' '*N=' ' is completely blank.

This, in combination with the other nonblank column headers, confuses PROC TABULATE. The result is that some of the grid characters do not print.

Avoiding this problem is simple—don't create columns with all blank headers. This is a good idea from a table design perspective anyway. The following code fixes the problem by allowing the N label to print:

```
PROC TABULATE DATA=TEMP F=8.;
   CLASS RACE EDUC AGE;
   VAR INCOME;
   TABLE AGE, ALL=' '*N
      (RACE=' ' EDUC)*INCOME*MEAN=' '
      / RTS=10;
RUN;
```

The corrected table is shown in Output 19.22. Now the entire table grid is printed.

There are a couple of other ways that you can trigger problems with the table grid in the row and column headings, though all are rare. If you follow these three rules, you will avoid most heading problems:

- Make sure each row and column heading has at least one nonblank label.
- Make sure there are no extra parentheses in the row or column definition.
- Use the ROW=FLOAT option.

Output 19.21

		White	Black	Other	Education		
					<HS	HS	College
		Income	Income	Income	Income	Income	Income
Age							
20-39	1916	28709	27128	30740	16712	23452	32454
40-59	1933	34320	37067	33978	18696	25043	41699
60+	335	33263	38485	37702	23012	25201	43151

Output 19.22

		White	Black	Other	Education		
					<HS	HS	College
	N	Income	Income	Income	Income	Income	Income
Age							
20-39	1916	28709	27128	30740	16712	23452	32454
40-59	1933	34320	37067	33978	18696	25043	41699
60+	335	33263	38485	37702	23012	25201	43151

FUZZ Option Does Not Work

Actually, the FUZZ option does work, it's just that it doesn't do what you think it does. The FUZZ option is not well documented in the PROC TABULATE manual, so you may be confused about how it works. The following example shows what goes on when you specify a FUZZ setting. The example uses a very simple data set, which is printed for your reference in Output 19.23.

```
PROC TABULATE DATA=TESTFUZZ;
    CLASS ROWCAT;
    VAR THEVALUE;
    TABLE ROWCAT, THEVALUE*(N MEAN STD SUM) / RTS=8;
RUN;
```

This code creates a table that does not use the FUZZ option. This table is shown in Output 19.24. Next, the table is run again with the FUZZ option in place, using the following TABLE statement:

```
TABLE ROWCAT, THEVALUE*(N MEAN STD SUM) / FUZZ=2 RTS=8;
```

The new table is shown in Output 19.25. As you can see, the two tables are quite different. The FUZZ option causes all cells with values of less than 2 to be displayed and treated as zeros. That explanation sounds simple enough until you look closely at the table and try to make sense of it. For example, if you look at the row for ROWCAT=C, you see that the values have an N of zero, a missing value for the MEAN, but a SUM of 4.

To understand what happened, we will go through the table cell by cell. Looking first at the column of N statistics, you can see that the values for A and B are unchanged. These cell values were both larger than the FUZZ value of 2. But the values for C and D have been set to zero because they were both less than or equal to 2.

In the next column, things get more confusing. In the table without the FUZZ option, you can see that the means for C and D used to be 2.0 and 1.0, respectively. Because both are less than or equal to 2, you would expect them to be displayed as zeros in the second table. Instead, they are displayed as missing! This is because to compute the mean, PROC TABULATE takes the sum of the variables and divides by the number of observations. In this example, the N for C and D have both been set to zero by the FUZZ option, so the mean cannot be calculated.

And just when you thought things could not get more confusing, we come to the third column, the SUM. In this column, the value for row C is unchanged. Even though the number of observations is displayed as zero, PROC TABULATE does not need to use the number of observations to compute a sum, so it goes ahead and sums up the values. Because the sum of 4 is greater than the FUZZ factor, the value is displayed. In row D, the sum is only 1, so it is displayed as a zero.

In the fourth column, the standard deviation for row B is shown as zero because the value of 1.73 is less than 2. The value for row C was already zero, so it does not change. And the value for row D also does not change because there is still only one observation, so no standard deviation can be calculated.

As you can see, these computations are confusing and are unlikely to be of any use in the real world. No end user would ever be able to figure out this table. The next example shows how to use the DATA step and PROC FORMAT to achieve the desired effect of screening out data with low values and observation counts.

Output 19.23

OBS	ROWCAT	THEVALUE
1	A	9
2	A	5
3	A	5
4	A	2
5	A	2
6	B	4
7	B	4
8	B	1
9	B	1
10	C	2
11	C	2
12	D	1

Output 19.24

	The Value			
	N	MEAN	STD	SUM
RowCat				
A	5.00	4.60	2.88	23.00
B	4.00	2.50	1.73	10.00
C	2.00	2.00	0.00	4.00
D	1.00	1.00	.	1.00

Output 19.25

	The Value			
	N	MEAN	STD	SUM
RowCat				
A	5.00	4.60	2.88	23.00
B	4.00	2.50	0.00	10.00
C	0.00	.	0.00	4.00
D	0.00	.	.	0.00

FUZZ Option Does Not Work - the Solution

If you want to eliminate small values from your table, there are several approaches that you can use instead of the FUZZ option. First, you have to figure out what you want to do.

If you are trying to get rid of observations with small values for your analysis variable, then you can do this in the DATA step. For example, if you want to eliminate records with incomes of less than $10,000 from your table, you can't do this with FUZZ=10000. Instead, you can use the following code:

```
DATA TEMPFUZZ;
   SET TEMP;
   IF INCOME<10000 THEN DELETE;
RUN;
```

Now, you can run your table on the new data set TEMPFUZZ, and your table will not include any observations with income of less than $10,000.

Another thing you might want to do is eliminate rows or columns from your table where the number of observations is less than 10. If you use FUZZ=10, this will not remove the row or column for the table. And, as we saw in the previous example, what it does to that row will depend on the statistic. The N will be displayed as 0, the MEAN will be displayed as missing, and the SUM may be displayed without any change.

Instead, what you want to do is preprocess the data before running PROC TABULATE. You can do this via a PROC SUMMARY and a DATA step. Let's say you want to create a table of income broken down by race and gender. But you don't want to display any values of RACE with fewer than 10 observations. The following code does the trick:

```
PROC SUMMARY DATA=TEMP NWAY;
   CLASS RACE;
   VAR INCOME;
   OUTPUT OUT=TEMPSUM N=NUMOBS;
RUN;
DATA TEMPSUM2;
   SET TEMPSUM;
   IF NUMOBS>10;
RUN;
PROC SORT DATA=TEMP; BY RACE; RUN;
DATA TEMP2;
   MERGE TEMP TEMPSUM2 (IN=HAS10);
   BY RACE;
   IF NOT HAS10 THEN DELETE;
RUN;
```

Now all values of RACE that have fewer than 10 total observations have been dropped from the data set TEMP2. You can use this data set to create your table.

The third way you might use the FUZZ option is to eliminate the display of extremely small table cell values. For example, you might want to delete means that are less than 1000 in your table. Unfortunately, if you use FUZZ=1000, it will affect all of the statistics in your table, not just the MEAN.

To create a fuzz factor of 1000, the best thing to do is create a format that you can apply to the mean statistic alone. The following code shows how to use a PICTURE format to recode all values less than 1000 to a special text string.

```
PROC FORMAT;
   PICTURE FUZZLOW  0-<1000='<1000'
                    1000-HIGH='00000000';
RUN;
PROC TABULATE DATA=TEMP;
   CLASS CLASSVAR;
   VAR NUMEMP;
   TABLE CLASSVAR, NUMEMP*(N*F=8. MEAN*F=FUZZLOW.);
RUN;
```

This code creates a format that displays integers of up to 8 digits for the N statistic, integers of up to 8 digits for means that are 1000 or higher, and the text '<1000' for means that are less than 1000.

When you want to apply a fuzz factor to your table, it is easier to use some combination of the three techniques presented here than it is to use the FUZZ option.

CHAPTER 20

Limitations of PROC TABULATE and How to Get Around Them

No one tool can do every job, and PROC TABULATE is no exception. There are some things that the procedure just can't do.

This chapter shows examples of tasks that PROC TABULATE was not designed to do.

For some of these tasks, there are ways you can trick PROC TABULATE into producing the desired output. This chapter includes examples and code to show how you can produce the desired result.

For other tasks, another SAS tool may be recommended. For example, PROC PRINT and PROC REPORT are also good tools to use for some types of tables.

Then there are tasks where the work you are doing is so complex that you may have to use a variety of SAS and other tools to bring together the desired elements into your table.

You will notice that some of the examples in this chapter apply only to Version 6 of SAS. Version 7 has new features that make some of the limitations presented in this chapter go away.

Quantile Statistics (Version 6)

PROC TABULATE uses the same list of statistics that are available for other SAS procedures like PROC MEANS, but it does not include the extensive list of statistics that you can get from PROC UNIVARIATE. Generally, these additional statistics are not needed for your tables, but there's one exception.

It is common to display median values for your analysis variable. For variables with skewed distributions, displaying only a mean can be misleading. Unfortunately, PROC TABULATE does not support medians or other quantile statistics, as you can see from the log file displayed in Output 20.1. The error message in this log shows what happens if you request a median from PROC TABULATE.

There is a trick you can use to add statistics that were not generated by PROC TABULATE to your table. What you have to do is compute the statistic using whatever procedure you need, and then merge the statistical results with the main data set before running PROC TABULATE.

The only SAS procedure (in Version 6) that can calculate medians is PROC UNIVARIATE. So to get the medians, we need to run UNIVARIATE and output the medians to a data set. Then the medians are merged with the original data. The trick to making this work is that you run PROC UNIVARIATE with your PROC TABULATE classification variables as BY variables. Because our table is categorized by AGE and GENDER, these are the BY variables to use with UNIVARIATE.

```
PROC SORT DATA=TEMP; BY AGE GENDER; RUN;
PROC UNIVARIATE DATA=TEMP NOPRINT;
   BY AGE GENDER;
   VAR INCOME;
   OUTPUT OUT=TEMPMED MEDIAN=MEDINCM;
RUN;
DATA TEMP2;
   MERGE TEMP TEMPMED;
   BY AGE GENDER;
RUN;
```

Next, the medians are merged back into the main data set. Now the main data set has a variable called MEDINCM that, for each observation, holds the median for that combination of AGE and GENDER. The only trick to using this new data set is to remember that because each observation in each combination of AGE and GENDER has the same value for MEDINCM, you can take the mean of MEDINCM and get the median for that group.

```
PROC TABULATE DATA=TEMP2;
   CLASS AGE GENDER;
   VAR INCOME MEDINCM;
   TABLE AGE*GENDER=' ',
      INCOME=' '*(N MEAN STD) MEDINCM=' '*MEAN='MEDIAN'
      / BOX='Household Income' ROW=FLOAT RTS=20;
RUN;
```

This code produces the table shown in Output 20.2. Notice how the column for the mean of MEDINCM is labeled as the median. This creates a table that looks like you used PROC TABULATE to compute a median.

Output 20.1

```
39    PROC TABULATE DATA=TEMP;
40          CLASS AGE GENDER;
41          VAR INCOME;
42          TABLE AGE*GENDER=' ', INCOME*(N MEAN STD MEDIAN);
43    RUN;


ERROR: The type of name (MEDIAN) is unknown.
NOTE: The SAS System stopped processing this step because of errors.
```

Output 20.2

```
--------------------------------------------------------------------
|Household Income |     N    |    MEAN   |    STD    |   MEDIAN   |
|-----------------+----------+-----------+-----------+----------- |
|Age     |        |          |           |           |           |
|--------+--------|          |           |           |           |
|20-39   |Female  |   1017.00|   34428.35|   22990.38|   31013.47 |
|        |--------+----------+-----------+-----------+----------- |
|        |Male    |    978.00|   22979.78|   15684.40|   21056.20 |
|--------+--------+----------+-----------+-----------+----------- |
|40-59   |Female  |    969.00|   45049.30|   30160.47|   39412.99 |
|        |--------+----------+-----------+-----------+----------- |
|        |Male    |   1002.00|   24417.46|   17912.07|   21217.87 |
|--------+--------+----------+-----------+-----------+----------- |
|60+     |Female  |    194.00|   43004.30|   39058.57|   31551.25 |
|        |--------+----------+-----------+-----------+----------- |
|        |Male    |    149.00|   22137.19|   16168.26|   19350.16 |
--------------------------------------------------------------------
```

In Version 7, quantile statistics have been added to PROC TABULATE, so the trick shown in this example is not necessary. See Chapter 24, "Version 7 Enhancements," for examples of quantile statistics under Version 7.

Advanced Statistical Tests

The previous example showed how to add a statistic not generated by PROC TABULATE to your table. You can use this approach to add more complex statistical test results to your table. For example, let's say that you have a table that shows mean income for males and females in a variety of occupations. What you want to know is whether the means vary within each occupation.

You can set up a table easily that shows the two means for each occupation. What you can't do is display the appropriate statistical test result. What you need is a column that shows the p-values for the F statistic from an ANOVA comparing the two means.

The way to do this is to use a DATA step trick to attach the F-test results to each row. Here's how it works. First, you run the statistical procedure and output the desired statistic to a data set. Then you merge the statistical results with the main data set using the CLASS variable values (gender and occupation) to link the results to the observations. Then you use PROC TABULATE to calculate the mean of the statistical result for each value of gender and occupation. Because the statistical result will be the same for every observation within each gender and occupation group, the mean will give us the result we need.

```
PROC ANOVA DATA=TEMP NOPRINT OUTSTAT=PVALS;
   BY OCCUP;
   CLASS GENDER;
   MODEL INCOME=GENDER;
RUN;

PROC SORT DATA=PVALS (KEEP=OCCUP PROB _TYPE_);
   WHERE _TYPE_='ANOVA';
   BY OCCUP;
RUN;
PROC PRINT DATA=PVALS; RUN;

DATA TEMP2;
   MERGE TEMP PVALS;
   BY OCCUP;
RUN;

PROC TABULATE DATA=TEMP2;
   CLASS GENDER OCCUP;
   VAR INCOME PROB;
   TABLE OCCUP=' ', GENDER=' '*INCOME=' '*MEAN*F=DOLLAR10.
      ALL=' '*PROB=' '*MEAN='P-value'*F=10.4
      / BOX='Income' ROW=FLOAT;
RUN;
```

You can see that the F-test results (shown in Output 20.3) match the results reported in the main table (shown in Output 20.4), proving that this approach has worked correctly.

This approach can be used to add a wide variety of statistics to the table. For additional examples, see Noga and Zhao (1997) in "References."

Output 20.3

OBS	OCCUP	_TYPE_	PROB
1	Managerial	ANOVA	0.0000
2	Professional	ANOVA	0.0000
3	Technical	ANOVA	0.0016
4	Sales	ANOVA	0.0000
5	Clerical	ANOVA	0.0000
6	Services	ANOVA	0.0000
7	Manufacturing	ANOVA	0.0000
8	Farming	ANOVA	0.0315

Output 20.4

Income	Male	Female	
	MEAN	MEAN	P-value
Managerial	$55,818	$32,128	0.0000
Professional	$50,055	$32,365	0.0000
Technical	$34,680	$25,386	0.0016
Sales	$42,982	$18,991	0.0000
Clerical	$31,298	$20,522	0.0000
Services	$25,563	$13,337	0.0000
Manufacturing	$29,223	$18,101	0.0000
Farming	$19,638	$9,761	0.0315

Printing Record Details

As we saw in the previous examples, it is possible to trick PROC TABULATE into doing almost anything. But sometimes it is better to use tools for their strengths. PROC TABULATE is good at summarizing and displaying data. It's not good at displaying individual records.

The following three examples show how to produce reports that display individual records. The same output is produced using three different tools: PROC TABULATE, PROC PRINT, and PROC REPORT.

```
PROC TABULATE DATA=DETAILS;
   CLASS FNAME LNAME YEAR;
   VAR EDUC INCOME NUMKIDS MARITAL;
   TABLE FNAME=' '*LNAME=' '*YEAR=' ',
      (INCOME*F=DOLLAR8. NUMKIDS*F=8. MARITAL*F=MARFT.)*MEAN=' ';
RUN;
```

This PROC TABULATE code is very contrived. PROC TABULATE was not designed to print individual records. To get a listing of individual records, the first three variables in the report are set up as CLASS variables and become row headings. This code isn't pretty, but it gets the job done. The resulting table is shown in Output 20.5.

```
PROC PRINT DATA=DETAILS NOOBS;
   BY FNAME LNAME;
   ID FNAME LNAME;
   VAR YEAR INCOME NUMKIDS MARITAL;
RUN;
```

Using PROC PRINT instead of PROC TABULATE is much simpler, and you get virtually the same output, minus the table grid. The results are shown in Output 20.6. The only problem with PROC PRINT is that you don't get much control over the format of the display. The next example uses PROC REPORT, which does give you this control. The results are shown in Output 20.7.

```
PROC REPORT DATA=DETAILS SPLIT='*' HEADLINE;
   COLUMN FNAME LNAME YEAR INCOME NUMKIDS MARITAL;
   DEFINE FNAME / ORDER LEFT "Name";
   DEFINE LNAME / ORDER LEFT " ";
   DEFINE YEAR / ORDER LEFT "Year";
   DEFINE INCOME / SUM RIGHT "Income";
   DEFINE NUMKIDS / DISPLAY RIGHT "No. of*Children";
   DEFINE MARITAL / DISPLAY LEFT "Marital*Status";
   BREAK AFTER FNAME / OL SKIP SUMMARIZE SUPPRESS;
RUN;
```

Though this PROC REPORT code is fairly simple, it gives you an idea of what you could do with PROC REPORT. You can control column order, width, spacing, and subtotals or totals. Also, the interactive version of PROC REPORT lets you build a report by manipulating objects in a WYSIWYG view.

Output 20.5

```
-------------------------------------------------------------------
|                            | INCOME |NUMKIDS|  MARITAL  |
|----------------------------+--------+-------+-----------|
|DONNA    |JONES   |1998     | $54,000|      3|DIV/WID/DEP|
|         |        |---------+--------+-------+-----------|
|         |        |1999     | $50,000|      4|MARRIED    |
|---------+--------+---------+--------+-------+-----------|
|JANE     |DOE     |1998     | $75,000|      0|MARRIED    |
|         |        |---------+--------+-------+-----------|
|         |        |1999     | $77,000|      0|MARRIED    |
|---------+--------+---------+--------+-------+-----------|
|TIM      |SAUNDERS|1998     | $15,000|      0|MARRIED    |
|         |        |---------+--------+-------+-----------|
|         |        |1999     | $16,000|      1|MARRIED    |
-------------------------------------------------------------------
```

Output 20.6

```
    FNAME          LNAME           YEAR    INCOME    NUMKIDS    MARITAL

    DONNA          JONES           1998    54000        3      DIV/WID/DEP
                                   1999    50000        4      MARRIED


    JANE           DOE             1998    75000        0      MARRIED
                                   1999    77000        0      MARRIED


    TIM            SAUNDERS        1998    15000        0      MARRIED
                                   1999    16000        1      MARRIED
```

Output 20.7

```
                                            No. of  Marital
Name                    Year      Income   Children  Status
-----------------------------------------------------------------
DONNA      JONES        1998       54000      3      DIV/WID/DEP
                        1999       50000      4      MARRIED
                                  ---------
                                   104000


JANE       DOE          1998       75000      0      MARRIED
                        1999       77000      0      MARRIED
                                  ---------
                                   152000


TIM        SAUNDERS     1998       15000      0      MARRIED
                        1999       16000      1      MARRIED
                                  ---------
                                    31000
```

Formatting Text Values

Another weakness of PROC TABULATE is that it does not give you much control over how your text row and column labels are displayed in the row and column headers. Unlike PROC PRINT and PROC REPORT, PROC TABULATE does not have a SPLIT= option. The following example shows what happens when you try to handle long labels in PROC TABULATE. The following code has long labels and values in the row headings, column headings, and the BOX label:

```
PROC TABULATE DATA=TEMP F=DOLLAR8.;
   CLASS AGE REGION;
   VAR INCOME;
   TABLE REGION='Plant Location'*AGE, ALL=' '*N='# Employees'*F=8.
      INCOME=' '*(MEAN="Mean' STD='Standard Deviation')
      / BOX='International Pharmaceutical Corp. Salaries';
RUN;
```

You can see in Output 20.8 that the row and column headings have several problems. The row heading for the 'Mid-Atlantic/Northeast' region does not fit in the space allowed, and PROC TABULATE has chosen to break up the label in an awkward fashion. Similarly, the column headings for the N and STD statistics have awkward word breaks.

To fix the problem, we could widen the row title space and the columns, but this would create a table that is larger than it needs to be. Also, to get a wide enough space for the values of REGION, we will be creating an equally wide space for the values of AGE. Because the age categories are fairly narrow, this creates a lot of wasted space in the row heading.

A better option would be to switch tools. PROC REPORT is much better at this kind of thing. You can give each column the space it needs and no more. You can also use the SPLIT= option to control where the headings break.

```
PROC REPORT DATA=TEMP SPLIT='*' HEADSKIP;
   COLUMN REGION AGE (INCOME=INCN INCOME=INCMEAN INCOME=INCSTD);
   DEFINE REGION / GROUP WIDTH=22 LEFT 'Plant Location';
   DEFINE AGE / GROUP CENTER 'Age Group';
   DEFINE INCN / N WIDTH=9 CENTER '#*Employees';
   DEFINE INCMEAN / MEAN FORMAT=DOLLAR8. CENTER 'Mean Income';
   DEFINE INCSTD / STD FORMAT=DOLLAR9. WIDTH=9 CENTER
      'Standard*Deviation';
   BREAK AFTER AGE / SKIP SUPPRESS;
   TITLE 'International Pharmaceutical Corp. Salaries';
RUN;
```

In this new table, shown in Output 20.9, you can see that each column has been given just the space it needs. The column for REGION is very wide so that the lengthy category descriptions will fit. The columns for AGE and the three statistics are narrower because the space is not needed. The split character '*' is used to create appropriate breaks in the labels for AGE and the standard deviation.

If we wanted, we could even add grid lines to this report so that it would look just like the PROC TABULATE table. To learn more about PROC REPORT, see Pass (1997) in the "References" chapter.

Output 20.8

```
 --------------------------------------------------
|International        |   #    |        |Standard |
|Pharmaceutical Corp.|Employe-|        |Deviati- |
|Salaries             |   es   |  Mean  |   on    |
|-------------------+--------+--------+-------- |
|Plant      |Age      |        |        |        |
|Location   |         |        |        |        |
|---------+---------|        |        |        |
|Mid-       |20-39    |   1596| $28,761| $20,131 |
|Atlantic- |---------+--------+--------+-------- |
|/Northea- |40-59    |        |        |        |
|st         |         |   1290| $34,197| $26,766 |
|---------+---------+--------+--------+-------- |
|Southeast|20-39    |     37| $34,646| $17,099 |
|         |---------+--------+--------+-------- |
|         |40-59    |     64| $42,727| $21,678 |
|---------+---------+--------+--------+-------- |
|Midwest   |20-39    |     79| $27,928| $18,138 |
|         |---------+--------+--------+-------- |
|         |40-59    |     92| $36,564| $19,020 |
|---------+---------+--------+--------+-------- |
|West      |20-39    |    101| $28,282| $15,614 |
|         |---------+--------+--------+-------- |
|         |40-59    |    195| $33,531| $18,852 |
 --------------------------------------------------
```

Output 20.9

```
International Pharmaceutical Corp. Salaries

                          Age       #      Mean    Standard
Plant Location            Group  Employees Income  Deviation

Mid-Atlantic/Northeast    20-39     1596   $28,761  $20,131

                          40-59     1290   $34,197  $26,766

Midwest                   20-39       79   $27,928  $18,138

                          40-59       92   $36,564  $19,020

Southeast                 20-39       37   $34,646  $17,099

                          40-59       64   $42,727  $21,678

West                      20-39      101   $28,282  $15,614

                          40-59      195   $33,531  $18,852
```

Complex Table Designs

Sometimes tables are just too complex to use PROC TABULATE or any other SAS procedure. You may need too many statistics or too much customization of the output. Don't worry, there are still other options. The DATA step is a powerful tool for data manipulation, and you can use this tool to create complex reports.

The following example is designed to whet your appetite for this type of report. It produces a table with statistics available only from PROC UNIVARIATE, and it gives you complete control over the format of the output. It even uses the power of the DATA step to annotate the results.

The code includes a PROC UNIVARIATE to compute the results, and then a DATA _NULL_ and a series of PUT statements to create the report. Notice how the DATA step is used to test the results for normality and produce a custom label depending on the outcome.

```
PROC UNIVARIATE DATA=TEMP NOPRINT;
   BY AGEGRP GENDER;
   VAR SCORE1 SCORE2 SCORE3;
   OUTPUT OUT=OUTSTATS N=N1 N2 N3
                       MEAN=MEAN1 MEAN2 MEAN3
                       STD=STD1 STD2 STD3
                       MEDIAN=MEDIAN1 MEDIAN2 MEDIAN3
                       NORMAL=NORMAL1 NORMAL2 NORMAL3;
RUN;
DATA _NULL_;
   SET OUTSTATS;
   IF _N_=1 THEN DO;
      PUT 'Table of Univariate Statistics';
      PUT '------------------------------------------------------------------';
      PUT @1 'Age' @8 'Gender' @16 'Measure'  @26 'N' @32 'Mean'
      @38 'Std Dev' @47 'Median' @55 'Normality Test';
      PUT '------------------------------------------------------------------';
      PUT ' ';
   END;
   FORMAT MEAN1-MEAN3 STD1-STD3 MEDIAN1-MEDIAN3 Z5.2 NORMAL1-NORMAL3 5.2;
   IF NORMAL1<.95 THEN NTEXT1='Not Normal'; ELSE NTEXT1='Normal';
   IF NORMAL2<.95 THEN NTEXT2='Not Normal'; ELSE NTEXT2='Normal';
   IF NORMAL3<.95 THEN NTEXT3='Not Normal'; ELSE NTEXT3='Normal';
   PUT @1 AGEGRP @8 GENDER @;
   PUT @16 'Score1' @24 N1 @32 MEAN1 @40 STD1 @48 MEDIAN1 @56 NORMAL1 @64 NTEXT1;
   PUT @16 'Score2' @24 N2 @32 MEAN2 @40 STD2 @48 MEDIAN2 @56 NORMAL2 @64 NTEXT2;
   PUT @16 'Score3' @24 N3 @32 MEAN3 @40 STD3 @48 MEDIAN3 @56 NORMAL3 @64 NTEXT3;
   PUT ' ';
RUN;
```

The report produced by this code is shown in Output 20.10. This is just a simple example of this technique. For more examples showing this type of reporting, see *Reporting From the Field* in the "References" chapter. Also, there have been numerous SUGI papers on DATA _NULL_ reporting tricks.

Output 20.10

```
Table of Univariate Statistics
-----------------------------------------------------------------
Age    Gender  Measure   N     Mean  Std Dev  Median  Normality Test
-----------------------------------------------------------------


20-39  Female  Score1   1017   10.08   28.91   10.34    0.98    Normal
               Score2   1017   10.10   09.66   09.91    0.99    Normal
               Score3   1017   10.88   48.77   10.05    0.99    Normal

20-39  Male    Score1    978   09.59   10.03   09.67    0.99    Normal
               Score2    978   09.66   09.92   09.51    0.99    Normal
               Score3    978   16.54   31.79   18.70    0.87    Not Normal

40-59  Female  Score1    969   09.04   30.51   09.01    0.98    Normal
               Score2    969   08.39   29.51   09.06    0.99    Normal
               Score3    969   12.88   50.74   14.52    0.99    Normal

40-59  Male    Score1   1002   10.26   09.84   10.14    0.99    Normal
               Score2   1002   09.35   30.17   10.10    0.99    Normal
               Score3   1002   20.62   07.79   19.21    0.89    Not Normal

60+    Female  Score1    194   09.76   29.69   09.28    0.98    Normal
               Score2    194   07.49   29.84   06.82    0.98    Normal
               Score3    194   09.65   46.32   08.74    0.98    Normal

60+    Male    Score1    149   09.56   10.48   08.98    0.98    Normal
               Score2    149   09.30   31.75   08.57    0.99    Normal
               Score3    149   19.62   07.03   18.10    0.86    Not Normal
```

Creating an Output Data Set - I (Version 6)

The first thing to know about creating output data sets from your PROC TABULATE output is that you can't. There is no option (prior to Version 7) that will turn your table output into a data set. That said, if your table is not too complicated, there may be a way to use other procedures to create a data set that is similar to your table output.

There are two approaches to creating output data sets. One is presented in this example, and the other is presented in the next example. Which one you choose will depend on what your table looks like.

Both examples will use the same starting table, based on the code below. This is a fairly typical table with a classification variable in each dimension and an analysis variable for the cells. It produces more than one statistic.

```
PROC TABULATE DATA=TEMP F=6.2;
    CLASS SECTOR OCCUP;
    VAR EMPYEARS;
    TABLE OCCUP,
        SECTOR*EMPYEARS*(N*F=6. MEAN STD MIN MAX)
        / RTS=12;
RUN;
```

The table is shown in Output 20.11. This table can be turned into a data set in two different ways. To use the first approach, you have to be willing to reorganize the table somewhat, as in the following code. The change is that all of the CLASS variables have been moved to the row dimension.

```
PROC TABULATE DATA=TEMP F=6.2;
    CLASS SECTOR OCCUP;
    VAR EMPYEARS;
    TABLE OCCUP*SECTOR,
        EMPYEARS*(N*F=6. MEAN STD MIN MAX)
        / RTS=28;
RUN;
```

If you are willing to have a data set based on this version of the table, shown in Output 20.12, then you can use the approach in the following code. If not, turn the page and look at your other option.

```
PROC SUMMARY DATA=TEMP NWAY;
    CLASS SECTOR OCCUP;
    VAR EMPYEARS;
    OUTPUT OUT=TABOUT N=N MEAN=MEAN STD=STD MIN=MIN MAX=MAX;
RUN;
PROC PRINT DATA=TABOUT (DROP=_TYPE_ _FREQ_); RUN;
```

What this code does is produce the same output with PROC SUMMARY. A printout of the data set is shown in Output 20.13. If you compare this data set to the table in Output 20.12, you can see that the data set matches the table output.

Output 20.11

	Sector								
	Private				Public				
	Yrs. Emp				Yrs. Emp				
	N	MEAN	STD	MIN	MAX	N	MEAN	STD	MIN	MAX
Occupation										
Managerial	1078	6.19	3.02	0.25	18.25	137	6.09	2.84	0.25	13.50
Professio-nal	1620	5.23	2.84	0.25	17.50	397	3.98	2.48	0.25	13.00
Technical	397	4.70	2.92	0.25	15.75	20	2.68	2.45	0.25	9.00
Sales	590	3.77	2.78	0.25	12.00	66	2.28	1.74	0.25	8.00

Output 20.12

| | | Yrs. Emp | | | | |
		N	MEAN	STD	MIN	MAX
Occupation	Sector					
Managerial	Private	1078	6.19	3.02	0.25	18.25
	Public	137	6.09	2.84	0.25	13.50
Professional	Private	1620	5.23	2.84	0.25	17.50
	Public	397	3.98	2.48	0.25	13.00
Technical	Private	397	4.70	2.92	0.25	15.75
	Public	20	2.68	2.45	0.25	9.00
Sales	Private	590	3.77	2.78	0.25	12.00
	Public	66	2.28	1.74	0.25	8.00

Output 20.13

OBS	SECTOR	OCCUP	N	MEAN	STD	MIN	MAX
1	Private	Managerial	1078	6.18576	3.01522	0.25	18.25
2	Private	Professional	1620	5.22608	2.83515	0.25	17.50
3	Private	Technical	397	4.69584	2.91735	0.25	15.75
4	Private	Sales	590	3.77034	2.77656	0.25	12.00
5	Public	Managerial	137	6.08577	2.84308	0.25	13.50
6	Public	Professional	397	3.97670	2.48155	0.25	13.00
7	Public	Technical	20	2.67500	2.44559	0.25	9.00
8	Public	Sales	66	2.28409	1.74306	0.25	8.00

Creating an Output Data Set - II (Version 6)

If you are not willing to reorganize your table to create the output data set, or if your table is more complex, then the following approach may work better. You can even use this approach with complex statistics like percentages.

The idea behind this method is that you read in the actual PROC TABULATE output to create your data set. This approach takes a bit more work but has a lot more flexibility than the PROC SUMMARY approach.

The first step is to output your PROC TABULATE table to a file. You can do this by using PROC PRINTTO, as in the following code. The other thing you need to do is get rid of all variable and statistic labels, the row dividers, and all but one of the FORMCHAR characters. This creates a simple text file, with commas between the table columns.

```
FILENAME LISFILE 'C:\TEMP\TABOUT.LIS';
PROC PRINTTO PRINT=LISFILE NEW; RUN;
PROC TABULATE DATA=TEMP F=6.2 NOSEPS FORMCHAR=',            ';
   CLASS SECTOR OCCUP;
   VAR EMPYEARS;
   TABLE OCCUP=' ',
      SECTOR=' '*EMPYEARS=' '*
      (N=' '*F=6. MEAN=' ' STD=' ' MIN=' ' MAX=' ')
      / RTS=14;
RUN;
PROC PRINTTO; RUN;
```

A printout of this text file is shown in Output 20.14. The printout is shown with lines above and below so that you can see the file structure. Notice that the actual table data does not start until the fourth line of the output file. To read the data in, you can use an INFILE statement. The FIRSTOBS and OBS options are used to select the appropriate four lines of output (lines 4-7).

```
DATA TABOUT;
   FORMAT OCCUP $12.;
   INFILE LISFILE DLM=',' FIRSTOBS=4 OBS=7 TRUNCOVER;
   INPUT OCCUP $ N1 MEAN1 STD1 MIN1 MAX1
      N2 MEAN2 STD2 MIN2 MAX2;
RUN;
PROC PRINT DATA=TABOUT; RUN;
```

The resulting data set has four observations, one for each value of occupation. There are two sets of statistics for each observation, one for SECTOR='Private' and one for SECTOR='Public.' As you can see from this simple example, this approach takes a lot of tweaking. Generally, creating an output data set by reading in your PROC TABULATE output is a lengthy trial-and-error process.

Output 20.14

,	,	Private				,	Public				,
,Managerial	,	1078,	6.19,	3.02,	0.25,	18.25,	137,	6.09,	2.84,	0.25,	13.50,
,Professional,		1620,	5.23,	2.84,	0.25,	17.50,	397,	3.98,	2.48,	0.25,	13.00,
,Technical	,	397,	4.70,	2.92,	0.25,	15.75,	20,	2.68,	2.45,	0.25,	9.00,
,Sales	,	590,	3.77,	2.78,	0.25,	12.00,	66,	2.28,	1.74,	0.25,	8.00,

Output 20.15

OBS	OCCUP	N1	MEAN1	STD1	MIN1	MAX1	N2	MEAN2	STD2	MIN2	MAX2
1	Managerial	1078	6.19	3.02	0.25	18.25	137	6.09	2.84	0.25	13.5
2	Professional	1620	5.23	2.84	0.25	17.50	397	3.98	2.48	0.25	13.0
3	Technical	397	4.70	2.92	0.25	15.75	20	2.68	2.45	0.25	9.0
4	Sales	590	3.77	2.78	0.25	12.00	66	2.28	1.74	0.25	8.0

If you are running Version 7 of SAS software, none of this effort is necessary. See Chapter 24 for a description of the OUT= output data set option.

PART 6

Special Topics

PROC TABULATE Options Reference

PROC TABULATE allows you to specify settings for a number of options that affect table output. This chapter explains each of the options, recommends appropriate default settings, and refers you to examples that show the effect of each option setting.

There are three types of options that affect PROC TABULATE. The first are options used in the PROC TABULATE statement. Next, there are a number of options that are included in the TABLE statement. Finally, there are system-level options that have an effect on PROC TABULATE output.

This chapter also has a section outlining the new options available in Version 7. For detailed information on how they work, see Chapter 24, "Version 7 Enhancements."

PROC TABULATE Statement Options

DATA=*name*

This one is not really an option. The DATA= setting tells SAS which data set to use to create the table. The data set you name can be a permanent data set, a temporary data set, or a SAS view.

You can leave off the data set name in your PROC TABULATE statement if you want to (SAS will use the most recently created data set), but it's a sloppy programming practice. If you don't include the data set name, it makes your code more confusing for other users. Also, if you rearrange the order of your procedures, the most recently created data set might change, leading to unintended consequences.

DEPTH=*number*

DEPTH refers to the maximum number of elements you can have in each table crossing. Most users will never need to change this option from its default setting of 10. There is no performance benefit to reducing the limit, and it takes a pretty complex table to require a setting greater than 10. To figure out the depth setting you need, count up the number of asterisks in the row dimension. Do the same for the column dimension and the page dimension (if applicable). Take the highest count (either row, column, or page) and add 1. As long as this number is less than 10, you do not need to use the DEPTH option.

FORMAT=*format*

The FORMAT option specifies the default format for numbers appearing in the cells of the table. It also controls the size of the cell. For example, if the value in the cell is 99.9 and the format is 4.1, the cell will be four characters wide — just large enough to hold the number. But if the format is 12.2, then value will still show as 99.9, but the cell will be 12 characters wide. You can use this feature to control column widths.

The default setting is BEST12.2. In general, you will want to pick a different setting. However, unless every cell in the table will have the same format, it is better to specify the formats individually for each variable or statistic. See Chapter 11, "Formatting Table Values," for examples showing the effect of the FORMAT option.

FORMCHAR='*string*'
FORMCHAR(index list)='*string*'

This option, which is also a system option, is used to specify the symbols that will be used for the borders of the table. The string you specify is a series of 11 characters. If you don't want to specify all 11 characters, you can use the index list to tell PROC TABULATE which characters you wish to change, but this is not recommended. It is a better documentation practice to specify all 11 characters every time. This makes your code more portable, if you want to reuse your tables in other programs. For more information on how to specify the string of characters, and recommended settings, see Chapter 12, "Modifying the Table Grid," and Chapter 22, "FORMCHAR Settings Reference."

MISSING

This option tells PROC TABULATE to treat missing values as a valid value of the CLASS variable. Instead of dropping observations with missing data on the CLASS variable, PROC TABULATE displays them as an additional row, column, or page. You should turn on this option during the table production phase. If you have no CLASS variables with missing data, then turning on the option does no harm. But if you do have missing data, and you didn't bother to try a test run with MISSING turned on, you may miss a major problem with your table. After you finish testing, you may decide to turn off MISSING for the final output. See "Missing Data: How to Find It" in Chapter 9, "Handling Missing Data," for more information on this option.

NOSEPS

This option tells PROC TABULATE to remove the row dividers in the table. This makes the table harder to read, but it saves a great deal of space. By taking out the row separators, you get twice as many rows on the page (the row separators take up a full line, just like the row values). If your table is fairly complex, you probably do not want to use this option.

ORDER=*order*

This option is used to specify the order of the CLASS variable values in your table. If you do not specify an order, PROC TABULATE uses ORDER=INTERNAL. This means that the CLASS variable categories are sorted by their values (0 comes before 1, A comes before B). This applies to the variable's unformatted values.

The most common reason to change this setting is if you want to force a particular order. You can use the PROC FORMAT to create formats for your variable that have different order. If you then set the option to ORDER=FORMATTED, PROC TABULATE ignores the actual values of the data and sorts based on the formatted values. See "Ordering the Headings" in Chapter 10, "Modifying Row and Column Headings," for an example that uses ORDER=FORMATTED to reorganize a table.

If you use ORDER=DATA, this tells PROC TABULATE to list the CLASS variables in the order that they are encountered in the data set. If your data is sorted in a particular order and you want that order for the table, use ORDER=DATA.

The ORDER=FREQ setting is used when you want the class values with the most observations to appear first. If you are building a lengthy table, this is a good way to draw attention to the most important categories.

VARDEF=*divisor*

This option refers to the statistical method used to compute variances. The divisor value tells PROC TABULATE which divisor to use in calculations. The default is VARDEF=DF or degrees of freedom. Other options are N, WDF, and WGT. It is unlikely that you will want to change this setting from the default.

Recommended PROC TABULATE Statement Option Settings

In general, a good starting setting for each of the options is the default setting. You do not need to specify these options in the PROC TABULATE statement. The only nondefault option that is recommended is the MISSING option.

PROC TABULATE DATA=*name* MISSING;

TABLE Statement Options

BOX='*string*'
BOX=*varname*
BOX=_PAGE_

The BOX setting refers to the box at the top left of the table grid (above the rows and to the left of the column headings). If you are creating a single-page table, you can take advantage of this space to include a title for the table. You have the option of typing in the exact title between quotes or using a variable name. If you use a variable name, PROC TABULATE displays the variable's label (if it has one) or the name of the variable. See "Removing a Heading by Putting It In the Table BOX" in Chapter 13 for an example of the BOX= option.

The BOX=_PAGE_ setting is a special setting that only applies to three-dimensional tables. BOX=_PAGE_ creates a label for the box that shows the name and value of the page dimension variable for each page. It is highly recommended that you use BOX=_PAGE_ any time you have a three-dimensional table. See "Putting a Title in Each Table" in Chapter 6, "Creating Three-Dimensional Tables," for an example of the BOX=_PAGE_ option.

CONDENSE

If you want to fit more information on a single page, try the CONDENSE option. If your table is short but wide and spans more than one page, CONDENSE will remove some of the page breaks and put the start of the table as well as the "continued" part of the table on the same page.

If you have a small three-dimensional table, CONDENSE will remove some of the page breaks and print more than one table on a page. See "Fitting Tables to the Page" in Chapter 6 for an example of this option.

If your table is too tall for two tables to fit on a single page, CONDENSE has no effect. Also, if your page breaks are caused by a BY statement, CONDENSE has no effect.

FUZZ=*number*

This option allows you to treat very small numbers as zero. See the two examples on FUZZ= in Chapter 19, "Incorrect Tables and How to Fix Them," for an explanation of how this setting affects your table. In general, this option will not give you the control you need to handle small values in your data set and table. Instead, try using a DATA step to handle small numbers in the data set, and then create a format to handle the small numbers in your table cells.

MISSTEXT='*string*'

This is a handy option for making your tables more readable. By default, SAS displays the missing value code '.' in any table cell with missing data. MISSTEXT allows you to specify a different text string (up to 20 characters). Some common strings to use: '-', 'N/A', 'none', 'unknown', or 'no data'. See "Formatting Missing Values" in Chapter 9 for an example of this option.

PRINTMISS

This option affects three-dimensional tables only. If you have missing data for one of the CLASS variables in a three-dimensional table, PRINTMISS forces the rows and columns on each page to be the same. On pages where one of the CLASS values has no data, PROC TABULATE displays that row or column anyway so that the tables look consistent from page to page. If the entire logical page contains only missing values, the page does not print, regardless of the PRINTMISS option. See "Three-Dimensional Tables with Missing Data" in Chapter 9 and "Handling CLASS Variables with Missing Data" in Chapter 15 for examples of this option.

ROW=*spacing*

This option tells PROC TABULATE whether to leave space for every element in the row crossing, even if their labels are blank. The default is ROW=CONSTANT, which results in extra grid lines and wasted space in the row headings for some tables. Generally, you should reset this option to ROW=FLOAT to make better use of your row heading space. See "Modifying Row Headings" in Chapter 10 for an example of this option.

RTS=*number*

By default, PROC TABULATE uses one-quarter of the page width to hold row headings and the remaining three-quarters of the page for table values. If you have a table with long or complex row headings, but fairly simple or narrow columns, you may want to allot more space for the row headings. RTS= allows you to set the number of spaces used for the row headings, so you can choose a larger number. Conversely, if you are trying to save space to get a large table to fit on a single page, you may wish to use RTS= to reduce the space used by the row headings.

For example, if you set RTS=60, then the row heading will be 60 spaces wide. If you have one row variable, it will get all 60 spaces. If you have two nested row variables, each variable will get 30 spaces for the heading. The space is always divided evenly between the variables in a nested heading, even if one variable requires much less space than another does. See "Modifying Row Heading Widths" in Chapter 10 and "Reducing the Space Used by Row Headings" in Chapter 13 for examples of this option.

Recommended TABLE Statement Option Settings

For one- and two-dimensional tables, start with the following settings. While you don't always need ROW=FLOAT, it does no harm to have it in place.

```
TABLE row specification, column specification / ROW=FLOAT;
```

For three-dimensional tables, start with the following settings:

```
TABLE row specification, column specification / ROW=FLOAT BOX=_PAGE_;
```

System Options that Affect PROC TABULATE Output

LINESIZE=_number_
PAGESIZE=_number_

You use these options to tell PROC TABULATE how much text it can print on each physical page. You will need to figure out the maximum length and width that your printer can handle. A handy way to set the LS and PS values is to create two OPTIONS statements at the top of your program, one for pages printed portrait, and one for pages printed landscape. Depending on the tables that you are printing, you can use an asterisk to comment out the options settings that you don't need. To figure out the maximum settings your printer can handle, use PROC PRINT to create some test output and then experiment with various settings.

FORMCHAR=_'string'_

This option is explained under PROC TABULATE statement options. It can also be applied as a system option. This is handy if you are producing a number of tables that will all have the same FORMCHAR setting.

MISSING=_'string'_

This option specifies what text to display for missing values. The default is a period. This system option can be used as an alternative to the MISSTEXT= option on the TABLE statement if you want all of your tables to use the same text for missing values.

CENTER|NOCENTER

This option allows you to specify how the output will be placed on the page. You have the option of centering the output (CENTER) or aligning the output on the left margin (NOCENTER). This setting is a matter of preference, unless you are planning to export your output to a spreadsheet or word processor. In that situation, the left-justified NOCENTER option is preferable. PROC TABULATE uses spaces, rather than tabs, to align its output, and these can cause problems in other applications. If you want to center your output in a spreadsheet or word processor, do so after importing the table from SAS. See Chapter 16, "Outputting the Results," for more information on exporting PROC TABULATE tables.

Recommended System Option Settings

As a starting point, try the following:

```
OPTIONS LINESIZE=60 PAGESIZE=79 NOCENTER;
```

Or, for landscape output, try:

```
OPTIONS LINESIZE=130 PAGESIZE=60 NOCENTER;
```

For maximum flexibility, go ahead and set up both OPTIONS statements at the top of your code, and comment out the one you don't need:

```
OPTIONS LINESIZE=60 PAGESIZE=79 NOCENTER; *PORTRAIT;
*OPTIONS LINESIZE=130 PAGESIZE=60 NOCENTER; *LANDSCAPE;
```

New Options in Version 7

Most of the changes to the TABULATE procedure in Version 7 involve new options. The biggest changes include more control over which values of your classification variables are included in the table, and options to control table styles using the new Output Delivery System (ODS). Each of these new options are summarized below, but are explained in more detail in Chapter 24.

New PROC TABULATE Statement Options

OUT=

This option allows you to specify an output data set for your PROC TABULATE results. This data set contains records for each combination of BY and CLASS variables in the table. This data set contains the following variables:

- the BY and CLASS variables
- a variable _TYPE_ that identifies which combination of the CLASS variables is being reported in that record
- a variable _PAGE_ to identify the value of the page variable
- a variable _TABLE_ that denotes the table number if there is more than one table
- variables for each of the statistics.

CLASSDATA=

If you would like to specify the valid combinations of categories of your classification variables to be used in the table, you can use this option to name a data set containing those categories. This option is most useful in combination with the next option, EXCLUSIVE.

EXCLUSIVE

This option specifies that only those combinations of classifications that are included in the CLASSDATA data set will be used in the table. This allows you to force the inclusion or exclusion of certain categories. This data set must contain all of the CLASS variables.

For example, if you want to build a table showing categories of gender and marital status, you could specify in advance that the groupings you want to see are 'female-single', 'male-single', 'female-married', 'male-married'. To do this, you would build a data set with variables for gender and marital status, and create four observations, one for each combination you want to see in the table. These categories will be used in the table, *whether or not they have any data*. Any other classifications will be ignored.

EXCLNPWGT

If you are running weighted analyses, this option excludes from your table any observations with zero or negative weight values. By default, observations with negative weights are treated like zero weights and are included in the total number of observations for your table.

NOTRAP

This option disables the trapping of overflow errors that occur during processing. Under certain operating systems (for example, Solaris), using this option substantially improves performance.

QMARKERS=, QMETHOD=, QNTLDEF=

These three options together control how quantile statistics are calculated. Unlike Version 6, you can produce quantile statistics like medians using the TABULATE procedure in Version 7.

Unless you are experiencing memory or performance problems, or you want to use a particular methodology, these options are probably best left alone. Consult the Version 7 documentation for more details on these options.

STYLE=

Specifies the style to use for the table cells when producing PROC TABULATE output. This option is used with ODS. See Chapter 24 for more details on this option.

New CLASS Statement Options

There are a number of new CLASS statement options. Because you may not want to apply the same options to all of your CLASS variables, PROC TABULATE now supports multiple CLASS statements. If necessary, you can use a separate CLASS statement for each classification variable and use different options for each variable.

PRELOADFMT

If you have assigned user-defined formats to your CLASS variables, this option uses these values to "preload" the row and column headings of your table. A heading is created for every possible value of your CLASS variables, whether or not this value appears in the data set. In other words, every category you have defined in your format appears in your table.

This option interacts with the PRINTMISS and EXCLUSIVE options. Without one of these options in place, PRELOADFMT has no effect. With PRELOADFMT and PRINTMISS, you get every possible combination of CLASS variable values in your table. With PRELOADFMT and EXCLUSIVE, you get only those values that are in the CLASS variables' formats.

EXCLUSIVE

This option, in combination with the PRELOADFMT, can be used to exclude from your table any values of the CLASS variables that are not included in your formats.

GROUPINTERNAL

If you are not using PRELOADFMT, this option saves some computer resources. It tells SAS not to apply formats to the CLASS variables during the sorting process that goes on prior to table production. This option has no effect on the table itself.

STYLE=

Specifies the style to use for the CLASS variable names and labels when producing PROC TABULATE output. This option is used with ODS. See Chapter 24 for more details on this option.

New CLASSLEV Statement

This statement is used with the Output Delivery System (ODS) and has only one option:

STYLE=

Specifies the style to use for the CLASS variable values when producing PROC TABULATE output. See Chapter 24 for more details on this option.

New KEYWORD Statement

This statement is used with the Output Delivery System and has only one option:

STYLE=

Specifies the style to use for the statistical keyword labels and the label for ALL when producing PROC TABULATE output. See Chapter 24 for more details on this option.

New TABLE Statement Option

STYLE=

Specifies the style to use for the table cells when producing PROC TABULATE output. This option is used with ODS. See Chapter 24 for more details on this option.

New VAR Statement Option

STYLE=

Specifies the style to use for the analysis variable name headings when producing PROC TABULATE output. This option is used with ODS. See Chapter 24 for more details on this option.

Options No Longer Supported

DEPTH=

This option is no longer needed under Version 7. PROC TABULATE can now determine the appropriate DEPTH setting.

Recommended Version 7 Option Settings

In general, you will not need to change your standard option settings when running your programs under Version 7. However, if you want to take advantage of new features, explore the new options. The following is a sample list of situations where you would gain from using the new options.

If you are using ODS to format your output, you will want to take advantage of the STYLE= options in the PROC TABULATE, CLASS, CLASSLEVEL, VAR, and TABLE statements.

If you are analyzing multiple-choice data, or any other data where you have predefined categories for all of your CLASS variables, you may want to use the CLASSDATA and EXCLUSIVE options to force all of the categories to appear in your table.

If there are only some CLASS variables where you have predefined categories, you can use a second CLASS statement and the PRELOADFMT and EXCLUSIVE options to force all of the categories to appear in your table. Similarly, if there are values for your classification variables that you want to exclude from your table, you can use the same approach. Exclude them from your format, and then use a CLASS statement with the PRELOADFMT and EXCLUSIVE options to remove these observations from the table. This can come in handy if you want to avoid having data errors cause out-of-range values to appear in your table.

FORMCHAR Settings Reference

Virtually all of the PROC TABULATE output presented in this book has been shown with the standard "| - - - - | + | - - -" FORMCHAR setting. This gives you tables bordered by dashes, bars, and plus signs. These border characters are designed to be compatible with all printers, even ancient line printers.

Unless you are using a line printer, you can make your output look a whole lot better if you use the "|- ⌐⌐ ╫╫ ⌐⌐" FORMCHAR string. This will generate a table where all of the cell, row, column, and table boundaries are smooth, connected lines.

So if this looks so much better, why does anyone use the old "| - - - - | + | - - -" setting? The reason is that the special characters used to create smooth table boundaries are *printer specific*.

Recent versions of SAS come with the default FORMCHAR setting for these smooth borders. Using this setting, your tables will look wonderful in the output window, but the borders may very well show up as strange characters when you print your table.

To create output with smooth table borders on your printer, you may have to open up your printer manual and figure out what characters to use for your printer.

CHAPTER 22

Testing the Default Settings

The first step in creating tables with smooth borders is to test the default setting ("|–┌┐ ┤├ └┘") that ships with more recent versions of SAS. You may be lucky enough to have the right printer for this default setting to work.

Try creating a sample table with the "|–┌┐ ┤├ └┘" setting. This setting may already be included in your default options or you may need to type it in again. You can change the setting by using an OPTIONS statement, or by including the FORMCHAR in the PROC TABULATE statement. Either way, the setting to try is as follows:

FORMCHAR='|–┌┐ ┤├ └┘'

Or, you can use the hexadecimal equivalent (easier to type):

FORMCHAR='82838485868788898A8B8C'X

Try running your table with this setting and send it to your printer. If the table does not look right, check the Fonts setting on the Options menu. It is probably set to Sasfont. Try switching to SAS Monospace or SAS Monospace Bold and try again to print your table.

If this still doesn't work, and you have more than one printer, try another printer. All you need to find is one printer where you can reliably print smooth-bordered tables. If this setting will not work on any of your printers, then go on to the next step.

Looking Up the Correct FORMCHAR Setting

At this point, you're going to have to read your printer manual. This is the only way that you will be able to find the appropriate characters to use for your FORMCHAR setting.

In most printer manuals there is a hexadecimal conversion table in the back that shows all printable characters and their corresponding hex representations. (Note: Many printers have multiple character sets, so make sure you know which character set is in use.)

What you are looking for is the 11 characters shown in Table 22.1. Use this table as a worksheet, and fill in the correct hex codes in the column provided.

Using the New FORMCHAR Setting

After you have found the hex equivalents for the 11 characters, list them (in the order shown in the table) as your FORMCHAR string. Follow the string with an X to indicate that this string is given in hexadecimal codes.

You should end up with something that looks something like the following examples. If you can't find this information for your printer, try using one of these two strings below. These are fairly common settings, so one might just match your printer.

FORMCHAR='BACDC9CBBBCCCEB9C8CABC'X

FORMCHAR='B3C4DAC2B1C3C5B4C0C1D9'X

Table 22.1

BORDER CHARACTER DESIRED	DESCRIPTION	HEX CODE (FILL THIS IN)
│	Vertical bar	
—	Horizontal bar	
┌	Upper left corner	
┬	Upper middle divider	
┐	Upper right corner	
├	Middle left divider	
┼	Middle middle divider	
┤	Middle right divider	
└	Lower left corner	
┴	Lower middle divider	
┘	Lower right corner	

Now try running a sample table with your new format character set. Even if the output looks wrong in the output window, try sending it to your printer. You should now have a printout with smooth table borders. If it doesn't work, don't panic. There are a few more things to try.

First, most printers have more than one character set. Try to figure out which one is in use. If you can't figure this out you can (1) call the printer manufacturer's technical support line and ask how to determine the current character set, or (2) try creating FORMCHAR settings for each of the possible character sets and see if one works.

If you think you are using the right character set and the printout still does not look right, then you may need to alter your SAS configuration file. Depending on your operating system, this file may list several character set options. Under Windows, you will see several optional '-wincharset' options in your config.sas file. Usually there are three possibilities, two of which will be commented out. Try using each of the other options and see if you can get your output to print correctly.

If all else fails, now is the time to beg help from other SAS users. First, try to find another user in your organization who uses the same model printer and has figured out the appropriate FORMCHAR string. If that doesn't work, you can try asking outside SAS users for help by posting a message to the SAS-L Usenet group on the Internet. You may be able to find someone else who has the same printer. (You should be able to find information about SAS-L on the SAS Web site at www.sas.com.)

Whatever you do, don't call SAS Technical Support. There is nothing they can do to help you because FORMCHAR settings are a hardware issue, not a SAS problem.

Creating Tables Using SAS/ASSIST® Software

If you're more of a point-and-clicker than a programmer, then SAS/ASSIST is the tool for you. You can build a PROC TABULATE table without having to write the PROC TABULATE code.

All you need to know to use SAS/ASSIST is how to identify your data set. The program will lead you through everything else.

Even if you're an experienced programmer, SAS/ASSIST still has something to offer. It's a quick and easy tool to use to build the basic structure of a table. You can then take the SAS/ASSIST output and modify the code to add enhancements to the table.

This chapter takes you step-by-step through the process of creating a table using SAS/ASSIST.[1,2]

---

[1] SAS/ASSIST is a separate module from base SAS software, and it may not be available for your system, or it may not have been installed.

[2] This example was created using SAS 6.12. The same features are available under Version 7, though the interface has been rearranged somewhat. Under releases prior to 6.12, some or all these features may not be available.

Step 1: Make a Rough Sketch of the Table

Figure 23.1

The first step is to make a sketch of the table you want to build. In this example, we build a simple two-dimensional table, with an analysis variable (age), a CLASS variable (occupation), four statistics, and a summary at the bottom of each column.

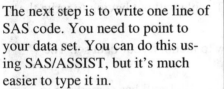

Step 2: Identify Your Data Set

Display 23.1

The next step is to write one line of SAS code. You need to point to your data set. You can do this using SAS/ASSIST, but it's much easier to type it in.

The syntax is to use a LIBNAME statement followed by the physical location of your data. This syntax is shown in the Program Editor window at the bottom of Display 23.1. Type in this line of code and submit it. Now you are ready to run SAS/ASSIST.

Step 3: Launch SAS/ASSIST

Display 23.2

To launch SAS/ASSIST, you need to select the Globals pull-down from the main menu bar, and then select SAS/ASSIST.[3]

Display 23.3

Now you should see the main menu of SAS/ASSIST, shown in Display 23.3. Next, select REPORT WRITING from the choices on the menu.

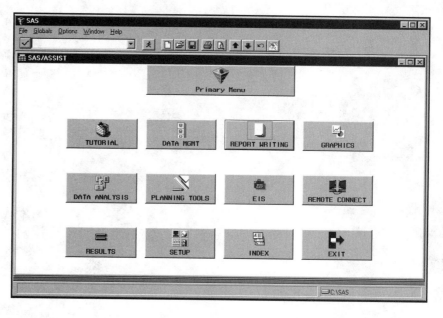

---

[3] You can also launch SAS/ASSIST by typing ASSIST in the command line just under the main menu bar.

Display 23.4

Now you should see the Report Writing Menu shown in Display 23.4. From this second menu, select TABLES.

Step 4: Pick a Table Design

Display 23.5

After clicking the TABLES button, you will see the menu of table styles shown in Display 23.5. This menu has two pages. The second page is shown in Display 23.6.

At this point, you can take the sketch of our table and find the style in Display 23.5 that seems to be the best fit. Our table design looks a lot like the Second style shown on the first page, so we'll choose that design.

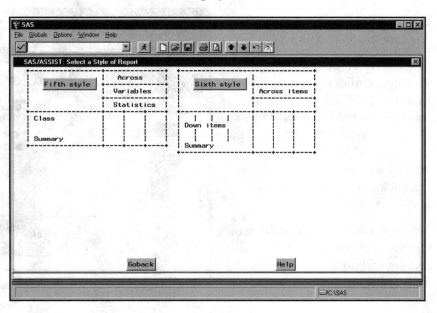

Display 23.6

Step 5: Selecting a Data Set

Display 23.7

This brings up a window that is set up just like our table design. It has a series of buttons on each part of the table to select the data set, variables, and statistics to display.

The first thing you have to do is to select the appropriate data set. To do this, click the first button on the page (titled `Active data set`).

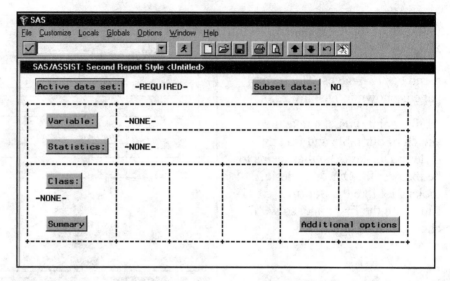

This will bring up a window where you can select the data set.

Because we set up our LIBNAME statement earlier to point to the location of our data set, that data set is now listed in the window that appears. To select the data set, you click on it. The data set is now marked with an asterisk, and you can click OK to confirm the selection.

This now takes us back to the main screen for the table. Notice that our data set will now be listed next to the `Active data set` button.

Display 23.8

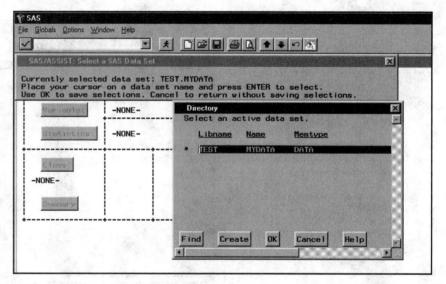

Step 6: Selecting an Analysis Variable

Now we'll move on to the next buttons on the page. Skip the Subset data button because we want to run this table on all of the data. The next button is for selecting the analysis variable.

When you click this button, a window opens up with a list of all of the numeric variables in the data set we just selected. In this case, you want to use AGE as the analysis variable, so click on AGE in this list. Again, an asterisk appears next to your selection, and you can click OK to confirm the selection.

Display 23.9

This takes you back to the main screen, and now AGE is shown in the column heading of the table.

The nice thing about SAS/ASSIST is that it is very visual. When you select the column variable, it is shown in the column heading. The same is true for row variables and statistics. You can see how your table is shaping up as you build it.

Display 23.10

Step 7: Selecting a CLASS Variable

Display 23.11

The next thing to do is select the row variable. As the table style indicates, this a CLASS variable that will be followed by a summary.

For our table, we want to show rows for each occupation, followed by a total. To select occupation as the row variable, click on the `Class` button. This brings up a list of all of the variables in the data set. You can then pick OCCUP from the list and click on OK.

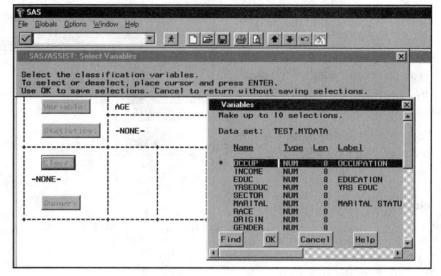

Display 23.12

When you get back to the main screen, OCCUP is now shown in the row header, indicating that it will be our row variable.

Step 8: Selecting the Statistics

Display 23.13

Next, we'll select the statistics for the table. Clicking on the Statistics button brings up a list of all of the statistics available for the analysis variable. This time, instead of picking one item from the list, we are going to pick four statistics by clicking on them. First, select the N statistic. Then you can select MEAN, MIN, and MAX.

As each statistic is selected, it moves to the top of the list and an asterisk is displayed. It is important that the statistics be selected in the order that you want them to appear in the table. If you had selected MAX, N, MEAN, and then MIN, the table would be built with the statistics in that order.

Display 23.14

After clicking OK, you come back to the main screen, and now the statistics are listed in the column header. This is a good time to check to be sure that all of the statistics are in the desired order.

Step 9: Requesting Column Totals

Display 23.15

The last thing we need to do is to add the totals at the bottom of each column. This is very simple: click on the Summary button, and then check the Summarize box in the window that appears next.

Display 23.16

After clicking OK, the main screen is now complete. We have selected a data set, analysis variable, classification variable, statistics, and a summary. Each of our selections is now shown in the display window.

Step 10: First Test Run

At this point, you are ready to try out the table specification. To run the table, you select Locals from the main menu bar, and the Run from the pull-down menu.

Display 23.17

As you have been selecting each of the options for your table in the previous steps, SAS/ASSIST has been busy behind the scenes writing the code to build your table. After you select Run, SAS/ASSIST submits the code. After the code has run, an output window appears with the desired table.

This table looks just like what we wanted. But if your table does not turn out right, close the output window, and go back to the main table screen and try again.

Display 23.18

		AGE			
	N	MEAN	MIN	MAX
OCCUPATION				
Managerial	770.00	43.33	25.00	85.00
Professional	865.00	41.50	25.00	77.00
Technical	144.00	38.65	25.00	71.00
Sales	447.00	42.90	25.00	75.00
Clerical	676.00	42.08	25.00	86.00
Services	530.00	41.57	25.00	78.00
Manufacturing	733.00	41.82	25.00	87.00
Farming	79.00	41.25	25.00	73.00
ALL	4244.00	42.03	25.00	87.00

Step 11: Reviewing the Code

Display 23.19

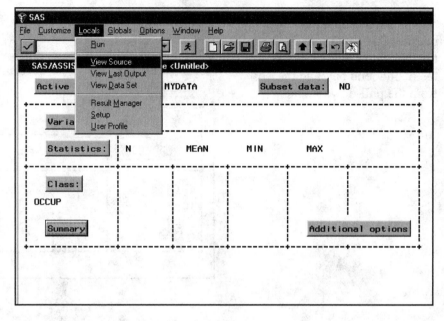

The best thing about SAS/ASSIST is that you can view and save the code it creates for your table. To view the code, select `Locals` from the main menu, and then `View source` from the pull-down.

Display 23.20

This brings up a special window that contains the SAS code used to create the table. Notice the detailed comments that SAS/ASSIST adds at the top of the code. These comments explain most of the options that were used and why they were selected.

Because this code is created as you select the options for your table, you can go look at it at any time. If you want to see how a particular option is coded, view the code before and after selecting the option to see what changed. You can save this code for future use. However, we're going to work with this table a bit more to improve its appearance before we save the code.

Step 12: Modifying the Table

Display 23.21

After closing the source code window, we are back at the main screen for our table. Now that we know the basic table works, we can turn to refining the table's appearance.

To do this, click the Additional Options button. This brings up a series of table options. One thing that would improve our table is modifying the formatting of the table values. We don't need to show two decimal places for our N statistic because it will always be an integer. By selecting the Format columns item, we can specify better formats.

Display 23.22

This brings up a window with a line for each column (one for each statistic). We can select more appropriate formats for each column. (8. for the N, MIN, and MAX, and 8.2 for the MEAN.)

Step 13: Second Test Run

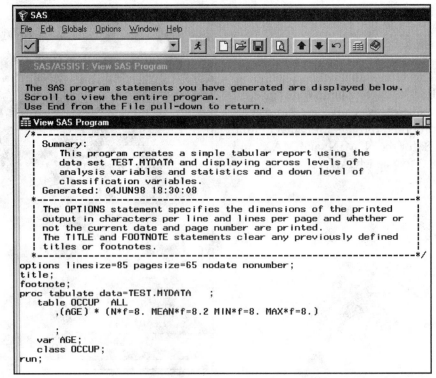

Display 23.23

After clicking OK twice to close the formatting and options windows, we're back at the main table screen. Now we can select Locals and Run again to look at our revised table.

As you can see in the output in Display 23.23, the table looks much better with the new formats in place.

Display 23.24

At this point, we are ready to view the code again. Close the output window, select Locals and then View Source to look at the code.

Step 14: Saving the Code

Display 23.25

This time we save the code. Do this by selecting File and then Save as to send the code to a file for future use.

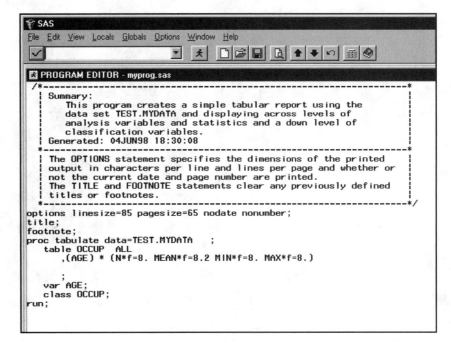

Now you can close SAS/ASSIST. To get out of SAS/ASSIST, close each of its windows.

This should take you back to your usual SAS setup. Now you can select File and then Open to select the code you just created and open it into the Program Editor. You can use this code to run the table anytime, without having to invoke SAS/ASSIST. You can also modify the code to add titles or footnotes, or use it as the basis for a more complex table.

Display 23.26

CHAPTER 24

Version 7 Enhancements

Version 7 is a major upgrade of the SAS System. It offers a number of new features that will enhance your PROC TABULATE output.

There are changes to base SAS software that make it easier to prepare your data. Also, there are a number of new features in PROC TABULATE. Finally, there is the Output Delivery System, which gives you much more control over the look and format of your output.

But if you don't need these new features, you will be relieved to hear that your existing DATA step and PROC TABULATE code is completely compatible with the new system. Every bit of SAS code in this book runs correctly under Version 7.

This chapter focuses only on those changes and enhancements that impact PROC TABULATE. To fully explore Version 7, review the *Changes and Enhancements* online documentation that comes with the software.

First, we look at the changes in base SAS software that affect PROC TABULATE. This includes several changes in the DATA step and naming conventions. Next, we look at changes in PROC TABULATE itself. Finally, we explore the new Output Delivery System and how it can be used to create more attractive tables.

Base SAS Changes: New Naming Conventions

One change in Version 7 is very obvious as soon as you look at the following example program:

```
PROC TABULATE DATA=TheWordyDataset;
   CLASS ReallyLongVariable ReallyLongVariableToo;
   VAR shorty;
   TABLE ReallyLongVariableToo,
      shorty*ReallyLongVariable=' '*MEAN=' '*F=18.
      / RTS=25 ROW=FLOAT;
RUN;
```

As you can see from this code and its results in Output 24.1, Version 7 now supports 32-character variable and data set names. Another new feature is mixed-case variable names. If you create a variable name with mixed upper- and lowercase, the variable is displayed in your output in the same case. Before, no matter how you typed the variable name in the DATA step, it was shown in uppercase in PROC TABULATE output.

You can get very descriptive in your SAS code when using Version 7. Not only is there support for longer data set and variable names, but you can also use longer variable and format labels–up to 256 characters. This is a great tool for producing easily understood code and output. However, for PROC TABULATE, these lengthy names and labels can be a problem. While you can create a bunch of 32-character variable names and give them each 256-character labels, this doesn't mean that you're going to have room to display these verbose labels in the row and column headings of a PROC TABULATE table. The following example shows the pitfalls of long names and labels:

```
PROC TABULATE DATA=TheWordyDataset;
   CLASS ReallyLongVariable ReallyLongVariableToo;
   VAR shorty;
   TABLE ReallyLongVariable*shorty,
      ReallyLongVariableToo*MEAN*F=6. / ROW=FLOAT;
RUN;
```

The resulting table is shown in Output 24.2. To fix this table, either you have to create extremely wide row and column headings, or you have to provide shorter labels in your PROC TABULATE code. The following code shows how to apply shorter labels to solve the problem:

```
PROC TABULATE DATA=TheWordyDataset;
   CLASS ReallyLongVariable ReallyLongVariableToo;
   VAR shorty;
   FORMAT ReallyLongVariable shortfmt.;
   TABLE ReallyLongVariable='Long'*shorty='Short',
      ReallyLongVariableToo='Long Too'*MEAN*F=6. / ROW=FLOAT;
RUN;
```

The revised table is shown in Output 24.3.

You may have noticed that Version 7 uses mixed case for statistic labels. In Output 24.2 and Output 24.3, the statistic keyword MEAN generates a column label of "Mean". All of the statistics now use mixed case for improved readability. If you want the old style labels, you'll have to apply them manually (MEAN="MEAN").

Output 24.1

```
-------------------------------------------------------------------
|                        |This is a short variable with a label |
|                        | so long it is completely ridiculous  |
|                        |--------------------------------------|
|                        |Widgets, Gadgets, |                   |
|                        |Whatchamacallits, |                   |
|                        | and Thigamagigs  |    Other Stuff     |
|------------------------+------------------+-------------------|
|ReallyLongVariableToo   |                  |                   |
|------------------------|                  |                   |
|2                       |             2860 |              355  |
|------------------------+------------------+-------------------|
|3                       |             1936 |              508  |
|------------------------+------------------+-------------------|
|4                       |             1310 |              181  |
|------------------------+------------------+-------------------|
|5                       |             1763 |               57  |
-------------------------------------------------------------------
```

Output 24.2

```
------------------------------------------------------------------
|                         |  ReallyLongVariableToo           |
|                         |----------------------------------|
|                         | 2   | 3   | 4   | 5    |
|                         |-----+-----+-----+------|
|                         | Mean| Mean| Mean| Mean |
|-------------------------+-----+-----+-----+------|
|ReallyLongVari-|         |     |     |     |      |
|able           |         |     |     |     |      |
|---------------+---------|     |     |     |      |
|Widgets,       |This is a short|    |    |    |   |
|Gadgets,       |variable with a|    |    |    |   |
|Whatchamacalli-|label so long  |    |    |    |   |
|ts, and        |it is          |    |    |    |   |
|Thigamagigs    |completely     |    |    |    |   |
|               |ridiculous     |2860|1936|1310|1763|
|---------------+---------------+----+----+----+----|
|Other Stuff    |This is a short|    |    |    |   |
|               |variable with a|    |    |    |   |
|               |label so long  |    |    |    |   |
|               |it is          |    |    |    |   |
|               |completely     |    |    |    |   |
|               |ridiculous     | 355| 508| 181|  57|
------------------------------------------------------------------
```

Output 24.3

```
-----------------------------------------------------------------
|                         |      Long Too                   |
|                         |---------------------------------|
|                         | 2   | 3   | 4   | 5    |
|                         |-----+-----+-----+------|
|                         | Mean| Mean| Mean| Mean |
|-------------------------+-----+-----+-----+------|
|Long           |         |     |     |     |      |
|---------------+---------|     |     |     |      |
|Widgets/Gadgets|Short    |2860 |1936 |1310 |1763  |
|---------------+---------+-----+-----+-----+------|
|Other Stuff    |Short    | 355 | 508 | 181 |  57  |
-----------------------------------------------------------------
```

Environment Changes: The Results Window

Other base SAS changes that affect PROC TABULATE are the improvements to the windowing environment. In addition to the Program, Log, and Output windows that are familiar to most SAS users, there is now a Results window.

The Results window holds a link to the output created from each of your runs since you started your current SAS session. You can easily review the output from your most recent submission, or from any code you submitted since launching SAS, as long as you don't clear the output window.

Display 24.1 shows an example of this window. Each type of SAS output is shown as a "folder," and each piece of procedure output (like a PROC TABULATE table) is shown as a "file." To view any of your previously generated output, you click on the appropriate procedure output icon in the Results window. This brings up the requested output in the Output window, as shown in Display 24.2.

This can be very handy when you are experimenting with PROC TABULATE table designs. You can set up your basic table; submit it, and then start playing with the table layout and options. A link to each version of the table is added to the Results window.

To compare version A to version B to version C, you can toggle back and forth quickly by clicking on the table icon in the output window. No more scrolling up and down through an endless output file.

Other Base SAS Changes

This is a brief preview of just a few of the new features in base SAS in Version 7. For more information, consult the printed or online documentation that comes with Version 7, or visit www.sas.com.

The rest of this chapter is devoted to more thorough coverage of changes in Version 7 that directly affect PROC TABULATE. This includes new PROC TABULATE statements, statistics, and options, as well as the new Output Delivery System.

Display 24.1

Display 24.2

Quantile Statistics: Medians

Until Version 7, if you wanted to display a median in a PROC TABULATE table, you had to cheat. You could add medians by using PROC UNIVARIATE and a DATA step, as shown in Chapter 20, "Limitations of PROC TABULATE and How to Get Around Them." Now it is easy to put a median in your table. MEDIAN has been added to the list of PROC TABULATE statistics in Version 7.

Under Version 6, if you submitted the following code, you'd get the error message shown in Output 24.4. This is because the keyword MEDIAN was not a Version 6 statistic.

```
PROC TABULATE DATA=TEMP;
   CLASS AGE GENDER;
   VAR INCOME;
   TABLE AGE*GENDER=' ', INCOME*(N MEAN MEDIAN);
RUN;
```

But submit the same code in Version 7 and you get the table shown in Output 24.5. Now you can compute a median without any hassle.

By default, PROC TABULATE computes the median using the same methodology as PROC UNIVARIATE. However, this method uses a lot of resources. If you are working with very large data sets and computer resources are an issue, you may want to use an alternate computational method.

Version 7 has a new option that allows you to select one of two computational methods for all quantile statistics. This option is called QMETHOD, and it is used in the PROC TABULATE statement. The default is QMETHOD=OS. The more efficient option is QMETHOD=HIST, which uses a different algorithm to approximate the median.

To see how this option affects the results, the following code shows the same table with the addition of the QMETHOD=HIST option. This data set has roughly 4,000 observations.

```
PROC TABULATE DATA=TEMP QMETHOD=HIST;
   CLASS AGE GENDER;
   VAR INCOME;
   TABLE AGE*GENDER=' ', INCOME*(N MEAN MEDIAN);
RUN;
```

The results are shown in Output 24.6. Notice how each of the medians is similar to the more exact calculations of the previous table, but are not identical.

The processing time and resources for the two runs in this example (one with the default QMETHOD=OS and one with QMETHOD=HIST) were virtually the same. You would have to have a much larger data set before there would be any payoff to using QMETHOD=HIST. Because using QMETHOD=HIST gives less accurate results, you would want to use it only when processing time is a major concern. For more information on the methodology involved in each QMETHOD, consult the Version 7 documentation.

Output 24.4

```
16    PROC TABULATE DATA=TEMP;
17         CLASS AGE GENDER;
18         VAR INCOME;
19         TABLE AGE*GENDER=' ', INCOME*(N MEAN MEDIAN);
10    RUN;

ERROR: The type of name (MEDIAN) is unknown.
NOTE: The SAS System stopped processing this step because of errors.
```

Output 24.5

```
-------------------------------------------------------------------------
|                              |                Income                  |
|                              |----------------------------------------|
|                              |    N     |    Mean     |   Median       |
|------------------------------+----------+-------------+------------     |
|Age            |              |          |             |                |
|---------------+--------------|          |             |                |
|20-39          |Female        |  1017.00 |   34428.35  |   30075.00     |
|               |--------------+----------+-------------+------------     |
|               |Male          |   978.00 |   22979.78  |   20800.00     |
|---------------+--------------+----------+-------------+------------     |
|40-59          |Female        |   969.00 |   45049.30  |   39500.00     |
|               |--------------+----------+-------------+------------     |
|               |Male          |  1002.00 |   24417.46  |   21181.50     |
|---------------+--------------+----------+-------------+------------     |
|60+            |Female        |   194.00 |   43004.30  |   30399.50     |
|               |--------------+----------+-------------+------------     |
|               |Male          |   149.00 |   22137.19  |   18634.00     |
-------------------------------------------------------------------------
```

Output 24.6

```
-------------------------------------------------------------------------
|                              |                Income                  |
|                              |----------------------------------------|
|                              |    N     |    Mean     |   Median       |
|------------------------------+----------+-------------+------------     |
|Age            |              |          |             |                |
|---------------+--------------|          |             |                |
|20-39          |Female        |  1017.00 |   34428.35  |   30600.58     |
|               |--------------+----------+-------------+------------     |
|               |Male          |   978.00 |   22979.78  |   20933.59     |
|---------------+--------------+----------+-------------+------------     |
|40-59          |Female        |   969.00 |   45049.30  |   39594.33     |
|               |--------------+----------+-------------+------------     |
|               |Male          |  1002.00 |   24417.46  |   21360.20     |
|---------------+--------------+----------+-------------+------------     |
|60+            |Female        |   194.00 |   43004.30  |   30476.42     |
|               |--------------+----------+-------------+------------     |
|               |Male          |   149.00 |   22137.19  |   19094.47     |
-------------------------------------------------------------------------
```

Quantile Statistics: Other Percentiles

In addition to medians, you can run any other quantile statistic under Version 7. All of the percentiles that used to be available only under PROC UNIVARIATE are now available to PROC TABULATE (and also PROC MEANS and PROC SUMMARY).

This includes the 1st, 5th, 10th, 25th, 50th, 75th, 90th, 95th, and 99th percentiles, as well as the interquartile range. The statistic keywords for these percentiles are P1, P5, P10, P25 (or Q1), P50 (or MEDIAN), P75 (or Q3), P90, P95, P99, and QRANGE, respectively.

What this means is that you can use PROC TABULATE to produce a UNIVARIATE-like listing of all of these percentiles and be able to package the output in a user-friendly table. The following code shows how this works:

```
PROC TABULATE DATA=TEMP;
   CLASS AGE GENDER;
   VAR INCOME;
   TABLE INCOME*F=DOLLAR9.*
      (MIN P1 P5 P10 P25 P50 P75 P90 P95 P99 MAX),
      AGE*GENDER=' '
      / RTS=20;
RUN;
```

This code specifies every possible percentile. The result, shown in Output 24.7, is an easy-to-read table. If this same information were produced using PROC UNIVARIATE, the output would span six pages and be much harder to read. You will still have to use PROC UNIVARIATE to compute some statistics (mode, skewness, kurtosis, and so on), but now you have more options for percentiles.

The second example shows the remaining statistical keywords available for the TABULATE procedure in Version 7. This version specifies Q1, MEDIAN, and Q3. These were also specified in the previous runs, using their alternate names: P25, P50, and P75. You can use whichever naming convention you like. This code also introduces one additional quantile statistic, the interquartile range. This statistic, specified as QRANGE, is the difference between the first and third quartiles (Q3-Q1).

```
PROC TABULATE DATA=TEMP;
   CLASS AGE GENDER;
   VAR INCOME;
   TABLE INCOME*F=DOLLAR9.*
      (Q1 MEDIAN Q3 QRANGE),
      AGE*GENDER=' '
      / RTS=20;
RUN;
```

The resulting table is shown in Output 24.8. Notice that Q1 in this table is indeed equivalent to P25 in the previous table. Similarly, MEDIAN matches P50 and Q3 matches P75.

By the way, you can use the QMETHOD option for all of these statistics. If you have a very large data set, this may save you computer resources, but it will reduce the accuracy of your estimates.

Output 24.7

		Age					
		20-39		40-59		60+	
		Female	Male	Female	Male	Female	Male
Income	Min	$0	$0	$0	$0	$0	$0
	P1	$0	$0	$0	$10	$0	$1
	P5	$5,431	$2,428	$8,000	$3,000	$8,742	$4,241
	P10	$10,050	$5,559	$14,098	$5,706	$12,722	$6,118
	P25	$19,500	$11,203	$25,000	$11,931	$18,693	$11,053
	P50	$30,075	$20,800	$39,500	$21,182	$30,400	$18,634
	P75	$45,020	$30,950	$57,000	$33,100	$51,174	$28,300
	P90	$62,000	$42,057	$90,669	$45,000	$99,421	$42,540
	P95	$79,137	$51,100	$101,999	$55,177	$116,000	$50,712
	P99	$102,402	$75,210	$146,320	$89,020	$224,999	$67,150
	Max	$219,998	$101,864	$228,492	$139,868	$251,998	$123,492

Output 24.8

		Age					
		20-39		40-59		60+	
		Female	Male	Female	Male	Female	Male
Income	Q1	$19,500	$11,203	$25,000	$11,931	$18,693	$11,053
	Median	$30,075	$20,800	$39,500	$21,182	$30,400	$18,634
	Q3	$45,020	$30,950	$57,000	$33,100	$51,174	$28,300
	QRange	$25,520	$19,747	$32,000	$21,169	$32,481	$17,247

CLASS Statement Options: PRELOADFMT

In Version 6, if one of the levels of your CLASS variable was not present in your data set, then you could not display it in a table. This is because before building a table, PROC TABULATE always checks your input data set and includes only those combinations of your CLASS variables that have data.

Even if you used a PROC FORMAT to define each of the categories of your CLASS variable, there was no way to get PROC TABULATE to use that information to create a table showing every category.

The following example illustrates the problem. In the code below, you can see that the variable GRADE is formatted into five categories. However, in the data set, there are no records for GRADE=1 ('D').

```
PROC FORMAT;
   VALUE GRADEFT 4='A'
                 3='B'
                 2='C'
                 1='D'
                 0='F';
RUN;
PROC TABULATE DATA=TEMP ORDER=FORMATTED;
   CLASS GRADE GENDER;
   FORMAT GRADE GRADEFT.;
   TABLE GRADE, N='Number of Students'*GENDER=' '*F=8.
      / RTS=10;
RUN;
```

As you can see in the resulting table, shown in Output 24.9, the row for GRADE='D' is left out. PROC TABULATE ignores this classification value, even though it is included in the user-defined format GRADEFT. In Version 6, there is nothing you can do about this except change your data set (see "DATA Step Tricks" in Chapter 9, "Handling Missing Data").

In Version 7, that has changed. You can now tell PROC TABULATE to use the format when building the table. In the sample code below, the PRELOADFMT option is used to tell PROC TABULATE to look at the formats before building the table. Notice that PRELOADFMT is used as a CLASS statement option. This is a new feature of Version 7, that you can now apply several options on the CLASS statement. In addition, the PRINTMISS option is used to tell PROC TABULATE to go ahead and display cells with missing data.

```
PROC TABULATE DATA=TEMP ORDER=FORMATTED;
   CLASS GRADE GENDER / PRELOADFMT;
   FORMAT GRADE GRADEFT.;
   TABLE GRADE, N='Number of Students'*GENDER=' '*F=8.
      / RTS=10 PRINTMISS;
RUN;
```

The resulting table is shown in Output 24.10. Now there is a row for GRADE='D', even though there is no data. The PRELOADFMT and PRINTMISS options together created the complete range of values.

Output 24.9

```
-----------------------------------
|          |     Number of        |
|          |     Students         |
|          |---------------------|
|          | Female  |  Male     |
|----------+---------+----------|
|Grade     |         |          |
|--------  |         |          |
|A         |    290  |    304   |
|----------+---------+----------|
|B         |    727  |    674   |
|----------+---------+----------|
|C         |    617  |    648   |
|----------+---------+----------|
|F         |    546  |    503   |
-----------------------------------
```

Output 24.10

```
-----------------------------------
|          |     Number of        |
|          |     Students         |
|          |---------------------|
|          | Female  |  Male     |
|----------+---------+----------|
|Grade     |         |          |
|--------  |         |          |
|A         |    290  |    304   |
|----------+---------+----------|
|B         |    727  |    674   |
|----------+---------+----------|
|C         |    617  |    648   |
|----------+---------+----------|
|D         |      .  |      .   |
|----------+---------+----------|
|F         |    546  |    503   |
-----------------------------------
```

 You do need to be careful when using the PRELOADFMT and PRINTMISS options together. PROC TABULATE will build a table with every possible combination of CLASS variable values, whether or not they make any sense.

CLASS Statement Options: EXCLUSIVE

The previous example shows how to use PRELOADFMT with PRINTMISS to force all possible values of your CLASS variables to be displayed in your table. You can also use PRELOADFMT to do the opposite. That is, you can use PRELOADFMT to limit your table to the values in your format.

You do this by pairing PRELOADFMT with the new EXCLUSIVE option. The EXCLUSIVE option tells PROC TABULATE to limit the table to classification values that are in your user-defined format and are also present in the data set.

For example, using the same data and table design as the previous example, you can see what happens when PRELOADFMT is used with EXCLUSIVE.

```
PROC TABULATE DATA=TEMP ORDER=FORMATTED;
   CLASS GRADE GENDER / PRELOADFMT EXCLUSIVE;
   FORMAT GRADE GRADEFT.;
   TABLE GRADE, N='Number of Students'*GENDER=' '*F=8. / RTS=10;
RUN;
```

As you can see in Output 24.11, the table is now limited to values of the classification variable that have non-missing data. The row for GRADE='D' is dropped from the table. In this example, using PRELOADFMT and EXCLUSIVE together produces the same result as not using PRELOADFMT at all. PROC TABULATE drops classification values with no data from the table.

So why have an EXCLUSIVE option if it doesn't do anything new? To see the reason, we need a different data set. In this case, we need a data set that has values that do not fit our format. The following code produces the same table, but this time the data set has changed somewhat:

```
PROC TABULATE DATA=OTHERDATA ORDER=FORMATTED;
   CLASS GRADE GENDER;
   FORMAT GRADE GRADEFT.;
   TABLE GRADE, N='Number of Students'*GENDER=' '*F=8. / RTS=10;
RUN;
```

As you can see in Output 24.12, we have two records where GRADE=7, which does not fit our format. This is where the EXCLUSIVE option comes in handy. We can use it to tell PROC TABULATE to ignore any values that are not part of our user-defined format.

```
PROC TABULATE DATA=OTHERDATA ORDER=FORMATTED;
   CLASS GRADE GENDER / PRELOADFMT EXCLUSIVE;
   FORMAT GRADE GRADEFT.;
   TABLE GRADE, N='Number of Students'*GENDER=' '*F=8. / RTS=10;
RUN;
```

Now, as shown in Output 24.13, the two records have been dropped from the table. Using the EXCLUSIVE option is a great way to deal with dirty data. Now you don't have to worry about stray values in your data turning up in your table. You can also use this as a way of subsetting your data, by building a format that only includes some of the values of your classification variable.

Output 24.11

| | | Number of Students | |
		Female	Male
Grade			
A		290	304
B		727	674
C		617	648
F		546	503

Output 24.12

| | | Number of Students | |
		Female	Male
Grade			
7		1	1
A		360	403
B		502	564
C		260	243
D		237	286
F		190	173

Output 24.13

| | | Number of Students | |
		Female	Male
Grade			
A		360	403
B		502	564
C		260	243
D		237	286
F		190	173

CLASS Statement Options: MISSING

The MISSING option itself is not new. It is already available as a PROC TABULATE statement option. What it does is treat missing values as a valid additional category for your classification variables. What's new is that you can now use this option in the CLASS statement.

To see how the MISSING option works in general, take a look at the following example program. Here the MISSING option is applied in the PROC TABULATE statement, so it affects every CLASS variable in the table.

```
PROC TABULATE DATA=TEMP ORDER=FORMATTED MISSING;
   CLASS GRADE RACE GENDER;
   TABLE GENDER=' '*GRADE,
      RACE=' '*N=' '*F=8.
      / BOX='Number of Students' RTS=20;
RUN;
```

The output from this code is shown in Output 24.14. Notice how there are additional rows in the table for GRADE=., and RACE=. is shown as an additional column.

This option is great if you want to show the missing category for every CLASS variable in your table, but what if you want to show the missing values for just one CLASS variable?

This is why the MISSING option is now available as a CLASS statement option. You can put a variable in its own CLASS statement and apply the MISSING option to that variable only.

The following code shows how to modify our table so that missing values for GRADE are shown, but missing values for GENDER and RACE are not:

```
PROC TABULATE DATA=TEMP ORDER=FORMATTED;
   CLASS RACE GENDER;
   CLASS GRADE / MISSING;
   TABLE GENDER=' '*GRADE,
      RACE=' '*N=' '*F=8.
      / BOX='Number of Students' RTS=20;
RUN;
```

In the resulting table, shown in Output 24.15, you can see that the missing data for GRADE is shown, but the missing data for RACE is now hidden. By using the MISSING option in the CLASS statement, you can decide variable by variable whether you wish to show missing data in your table.

Output 24.14

```
--------------------------------------------------------
|Number of Students|   .   | Black | Other | White |
|------------------+-------+-------+-------+-------|
|FEMALE  |GRADE    |       |       |       |       |
|        |---------|       |       |       |       |
|        |.        |     .|     .|     .|    47|
|        |---------+-------+-------+-------+-------|
|        |A        |     .|    29|     7|   278|
|        |---------+-------+-------+-------+-------|
|        |B        |     .|    37|    14|   451|
|        |---------+-------+-------+-------+-------|
|        |C        |     8|    45|    44|   163|
|        |---------+-------+-------+-------+-------|
|        |D        |     7|    18|    43|   169|
|        |---------+-------+-------+-------+-------|
|        |F        |    13|    15|    34|   128|
|--------+---------+-------+-------+-------+-------|
|MALE    |.        |     .|     .|     .|    72|
|        |---------+-------+-------+-------+-------|
|        |A        |     .|    18|    10|   304|
|        |---------+-------+-------+-------+-------|
|        |B        |     2|    43|    16|   503|
|        |---------+-------+-------+-------+-------|
|        |C        |     2|    48|    20|   173|
|        |---------+-------+-------+-------+-------|
|        |D        |     8|    22|    40|   216|
|        |---------+-------+-------+-------+-------|
|        |F        |     9|    13|    25|   126|
--------------------------------------------------------
```

Output 24.15

```
----------------------------------------------------
|Number of Students| Black | Other | White |
|------------------+-------+-------+-------|
|FEMALE  |GRADE    |       |       |       |
|        |---------|       |       |       |
|        |.        |     .|     .|    47|
|        |---------+-------+-------+-------|
|        |A        |    29|     7|   278|
|        |---------+-------+-------+-------|
|        |B        |    37|    14|   451|
|        |---------+-------+-------+-------|
|        |C        |    45|    44|   163|
|        |---------+-------+-------+-------|
|        |D        |    18|    43|   169|
|        |---------+-------+-------+-------|
|        |F        |    15|    34|   128|
|--------+---------+-------+-------+-------|
|MALE    |.        |     .|     .|    72|
|        |---------+-------+-------+-------|
|        |A        |    18|    10|   304|
|        |---------+-------+-------+-------|
|        |B        |    43|    16|   503|
|        |---------+-------+-------+-------|
|        |C        |    48|    20|   173|
|        |---------+-------+-------+-------|
|        |D        |    22|    40|   216|
|        |---------+-------+-------+-------|
|        |F        |    13|    25|   126|
----------------------------------------------------
```

CLASS Statement Options: ORDER

Just like the MISSING option, the ORDER option itself is not new. It is already available as a PROC TABULATE statement option. This option allows you to specify the order in which your CLASS variable categories are listed in your table. What's new is that you can now use this option in the CLASS statement.

To see how the ORDER option works in general, take a look at the following example program. Here the ORDER option is applied in the PROC TABULATE statement, so it affects every CLASS variable in the table. In this case, the ORDER=FREQ option has been chosen so that the values will be listed in descending order of their *frequency* in the data set.

```
PROC TABULATE DATA=TEMP ORDER=FREQ;
   CLASS GENDER AGE ANS;
   TABLE AGE*ANS=' ',
      (GENDER=' ' ALL)*N=' '*F=10.
      / RTS=20 ROW=FLOAT
      BOX='Reason for Choice';
RUN;
```

Reason for Choice (?)

The output from this code is shown in Output 24.16. Notice how the categories of the variable ANS are listed in order of their frequency. This makes it easier to see the most popular choices.

However, we also have the unintended effect of listing the age categories out of order. This is because the 40-59 age group has more observations than the 20-39 or 60+ groups.

This is why the ORDER option is now available as a CLASS statement option. You can put a variable in its own CLASS statement, and apply the ORDER option to that variable only.

The following code shows how to modify our table so that the variable ANS is ordered by frequency, but the variables GENDER and AGE are ordered by their formatted values:

```
PROC TABULATE DATA=TEMP ORDER=FORMATTED;
   CLASS GENDER AGE;
   CLASS ANS / ORDER=FREQ;
   TABLE AGE*ANS=' ',
      (GENDER=' ' ALL)*N=' '*F=10.
      / RTS=20 ROW=FLOAT
      BOX='Reason for Choice';
RUN;
```

In the resulting table, shown in Output 24.17, you can see that ANS is still ordered by frequency, but now the values of AGE appear in the proper order.

Output 24.16

```
-----------------------------------------------------------
|Reason for Choice | Female  |  Male   |   All    |
|------------------+---------+---------+--------- |
|Age     |         |         |         |          |
|--------+---------|         |         |          |
|40-59   |Location |     707 |     743 |    1450  |
|        |---------+---------+---------+--------- |
|        |Hours    |      32 |      28 |      60  |
|        |---------+---------+---------+--------- |
|        |Prices   |      19 |      14 |      33  |
|--------+---------+---------+---------+--------- |
|60+     |Location |     718 |     674 |    1392  |
|        |---------+---------+---------+--------- |
|        |Hours    |      28 |      23 |      51  |
|        |---------+---------+---------+--------- |
|        |Prices   |      21 |      25 |      46  |
|--------+---------+---------+---------+--------- |
|20-39   |Location |     167 |     136 |     303  |
|        |---------+---------+---------+--------- |
|        |Hours    |       9 |       4 |      13  |
|        |---------+---------+---------+--------- |
|        |Prices   |       2 |       6 |       8  |
-----------------------------------------------------------
```

Output 24.17

```
-----------------------------------------------------------
|Reason for        | FEMALE  |  MALE   |   All    |
|------------------+---------+---------+--------- |
|AGE     |         |         |         |          |
|--------+---------|         |         |          |
|20-39   |Location |     167 |     136 |     303  |
|        |---------+---------+---------+--------- |
|        |Hours    |       9 |       4 |      13  |
|        |---------+---------+---------+--------- |
|        |Prices   |       2 |       6 |       8  |
|--------+---------+---------+---------+--------- |
|40-59   |Location |     707 |     743 |    1450  |
|        |---------+---------+---------+--------- |
|        |Hours    |      32 |      28 |      60  |
|        |---------+---------+---------+--------- |
|        |Prices   |      19 |      14 |      33  |
|--------+---------+---------+---------+--------- |
|60+     |Location |     718 |     674 |    1392  |
|        |---------+---------+---------+--------- |
|        |Hours    |      28 |      23 |      51  |
|        |---------+---------+---------+--------- |
|        |Prices   |      21 |      25 |      46  |
-----------------------------------------------------------
```

CLASS Statement Options: DESCENDING, ASCENDING

Not only can you pick a different ORDER for each CLASS variable in your table, but you can also reverse that order if needed. ASCENDING and DESCENDING are new CLASS statement options that let you reverse the order of your CLASS variable values in the table.

To see why this might be useful, take a look at the following example. This code shows the Version 6 approach of using ORDER=FORMATTED to create a table with classification values ordered by their formatted values.

```
PROC TABULATE DATA=TEMP ORDER=FORMATTED;
   CLASS AGE GENDER;
   VAR INCOME;
   TABLE GENDER=' ', AGE*INCOME=' '*MEAN=' '
      / BOX='Mean Income' RTS=10;
RUN;
```

In Output 24.18, you can see that the values are listed in order of their formatted values (females before males, and the ages in numerical order). This was all done under Version 6. It works because the order of the formatted values was the order we wanted for our table, so ORDER=FORMATTED did the trick. But what if we want to list the values in a different order? Let's say we'd rather have a table that lists the incomes for males and then the incomes for females.

Under Version 7, not only can you specify an ORDER, but you can also use a DESCENDING or ASCENDING option to reverse the order. The following code shows the same table with the same ORDER option, but the addition of the DESCENDING option:

```
PROC TABULATE DATA=TEMP ORDER=FORMATTED;
   CLASS AGE GENDER / DESCENDING;
   VAR INCOME;
   TABLE GENDER=' ', AGE*INCOME=' '*MEAN=' '
      / BOX='Mean Income' RTS=10;
RUN;
```

The new output is shown in Output 24.19. Notice that now the order is males and then females, and the ages are now in reverse numerical order. This is a problem. While we might want to list males before females, we probably don't want to reverse the order of the ages.

Don't worry, this can be fixed. Another new feature in Version 7 is the ability to use more than one CLASS statement. We can use one CLASS statement for AGE and a second CLASS statement for GENDER. Then we can apply the DESCENDING option to the second CLASS statement only, as in the following example:

```
CLASS AGE;
CLASS GENDER / DESCENDING;
```

These CLASS statements generate the table shown in Output 24.20. Now males are again listed before females, but the ages are shown in the proper order.

Output 24.18

```
-----------------------------------------------------
|Mean    |                   Age                     |
|Income  |-------------------------------------------|
|        |    20-39    |    40-59    |    60+        |
|--------+-------------+-------------+---------------|
|Female  |    34428.35 |    45049.30 |    43004.30   |
|--------+-------------+-------------+---------------|
|Male    |    22979.78 |    24417.46 |    22137.19   |
-----------------------------------------------------
```

Output 24.19

```
-----------------------------------------------------
|Mean    |                   Age                     |
|Income  |-------------------------------------------|
|        |    60+      |    40-59    |    20-39      |
|--------+-------------+-------------+---------------|
|Male    |    22137.19 |    24417.46 |    22979.78   |
|--------+-------------+-------------+---------------|
|Female  |    43004.30 |    45049.30 |    34428.35   |
-----------------------------------------------------
```

Output 24.20

```
-----------------------------------------------------
|Mean    |                   Age                     |
|Income  |-------------------------------------------|
|        |    20-39    |    40-59    |    60+        |
|--------+-------------+-------------+---------------|
|Male    |    22979.78 |    24417.46 |    22137.19   |
|--------+-------------+-------------+---------------|
|Female  |    34428.35 |    45049.30 |    43004.30   |
-----------------------------------------------------
```

The DESCENDING statement is provided so that you can reverse the sort order when you have chosen ORDER=INTERNAL, FORMATTED, or DATA. If you have chosen ORDER=FREQ, the default is to list the data in descending order, so a new ASCENDING option has been added to reverse the order.

CLASSDATA Option

If you still can't get the exact table you want, even with all of these possible CLASS statement options, then there is one more new feature for you to try. The new CLASSDATA option allows you to create a data set that contains the exact combinations of classification variables that you want included in your table. To see why this option might be useful, take a look at the following table:

```
PROC TABULATE DATA=TEMP;
   CLASS GENDER RACE MARITAL;
   TABLE MARITAL, RACE=' '*GENDER=' '*N=' '*F=8.
      / RTS=20 BOX='Number of Observations';
RUN;
```

In Output 24.21, you can see that this table needs a little work. First, there is no data for black females. We could fix this by using the PRELOADFMT and PRINTMISS options to add a column for the missing data. But what if we want to make some other changes as well? For example, let's say we're only interested in two categories of marital status (Married and Divorced/Sep) and two categories of race (White and Black). We can make all of these changes at once by using a data set to show which values we want to include in the table. The following DATA step sets up our desired combinations of CLASS variables. The data set contains each of our CLASS variables, and the observations show each of the combinations of these variables that we want to see in the table. In this case, we want to see both values of GENDER, but only values 1 and 2 for RACE, and only values 1 and 3 for MARITAL.

```
DATA CLASSES;
   FORMAT GENDER GENDERFT. RACE RACEFT. MARITAL MARFT.;
   INPUT GENDER RACE MARITAL;
CARDS;
1 1 1
1 1 3
1 2 1
1 2 3
2 1 1
2 1 3
2 2 1
2 2 3
;
RUN;
```

The CLASSES data set explicitly lists the exact combinations of classifications that we want in our table. Now, these values will appear in the table, regardless of what values are present in the main data set TEMP. All we have to do is add the CLASSDATA=CLASSES option to the PROC TABULATE statement and the EXCLUSIVE option to limit our table to the values in data set CLASSES. This code produces Output 24.22.

```
PROC TABULATE DATA=TEMP CLASSDATA=CLASSES EXCLUSIVE;
   CLASS GENDER RACE MARITAL;
   TABLE MARITAL, RACE=' '*GENDER=' '*N=' '*F=8.
      / RTS=20 BOX='Number of Observations';
RUN;
```

Output 24.21

Number of Observations	White		Black	Other	
	Male	Female	Male	Male	Female
Marital Status					
Married	1214	1106	122	184	117
Widowed	8	63	.	1	8
Divorced/Sep	159	269	16	23	28
Never Married	298	287	40	47	35

Output 24.22

Number of Observations	White		Black	
	Male	Female	Male	Female
Marital Status				
Married	1214	1106	122	.
Divorced/Sep	159	269	16	.

When using this option, the variables in your CLASSDATA data set must have the exact names, lengths, and assigned formats as the variables in your table in order for the option to work. Notice how the example code uses a FORMAT statement in the DATA step that builds the CLASSES data set to assign formats to each of the variables. This was done so that the variables would match the main data set.

Other New Options

There are several other new options under Version 7 that are worth a brief mention. One new feature is the ability to use the WEIGHT option to apply weights specifically to individual variables. Just as the MISSING and ORDER options can now be applied to individual variables by using multiple CLASS statements, you can now apply weights to individual variables by using multiple VAR statements. Version 7 supports multiple VAR statements, and you can now use the WEIGHT option in the VAR statement.

This means that you can produce tables where different weights are used for different variables, or one variable is weighted and another is not. Or, by making a copy of your analysis variable, you can show that variable both weighted and unweighted in the same table. The following code shows how to do this.

Variable INCOME2 is just a copy of variable INCOME. It was created in a DATA step (not shown) using the statement INCOME2=INCOME. This allows us to display the unweighted results using INCOME and the weighted results using INCOME2. The WEIGHT=THEWT option in the second VAR statement tells PROC TABULATE to weight the variable INCOME2 using the weight contained in the variable THEWT. The resulting table is shown in Output 24.23. Notice how the numbers vary somewhat between the unweighted and weighted sections of the table.

```
PROC TABULATE DATA=TEMPWT;
   CLASS OCCUP GENDER;
   VAR INCOME;
   VAR INCOME2 / WEIGHT=THEWT;
   TABLE (INCOME='Unweighted' INCOME2='Weighted')*
      OCCUP=' '*MEAN=' '*F=DOLLAR10., GENDER=' ' ALL
      / RTS=30 BOX='Mean Income' ROW=FLOAT;
RUN;
```

Another new option also applies to weighted analyses. The PROC TABULATE statement option EXCLNPWGT will exclude from your table any observations with zero or negative weight values. By default, observations with negative weights are treated like zero weights and included in the total number of observations for your table. This option was added to permit uniformity between PROC GLM, PROC REPORT, PROC SUMMARY, PROC MEANS, and PROC TABULATE. By default, PROC GLM and PROC REPORT act as if the EXCLNPWGT option was used. Now you can use the option to create consistent output from PROC MEANS, PROC SUMMARY, and PROC TABULATE.

Two other new options affect how SAS uses computer resources. Neither has any effect on your output, but they can improve efficiency in certain situations. First, there is a new CLASS statement option called GROUPINTERNAL. This option specifies not to apply formats to the CLASS variables when PROC TABULATE preprocesses the data. This option saves computer resources when the CLASS variables contain discrete numeric values.

The second option is a PROC TABULATE statement option called NOTRAP. This option disables the trapping of overflow errors that occur during processing. In operating environments where the overhead of overflow error handling is significant, disabling the trapping for stable PROC TABULATE programs can considerably improve performance. Unless you are working with gigantic data sets, are experiencing extremely slow processing speeds, or are running out of available memory, you probably don't need to worry about the GROUPINTERNAL or NOTRAP options.

Output 24.23

Mean Income		Female	Male	All
Unweighted	Managerial	$47,400	$22,926	$34,570
	Professional	$38,626	$25,774	$31,879
	Technical	$28,601	$19,676	$24,235
	Sales	$38,299	$18,845	$31,567
Weighted	Managerial	$47,497	$23,060	$35,016
	Professional	$39,069	$25,985	$32,349
	Technical	$28,573	$20,018	$24,517
	Sales	$38,661	$19,489	$32,252

Output Data Sets

The good news is that Version 7 supports the creation of output data sets from your PROC TABULATE table. Even better news is that they are easy to specify. The bad news is that these data sets can be a little tricky to understand. PROC TABULATE tables can be quite complex, and these complexities are hard to represent in a simple data set. The following code illustrates a simple PROC TABULATE table with an output data set:

```
PROC TABULATE DATA=TEMP OUT=TABOUT;
   CLASS GENDER;
   VAR INCOME;
   TABLE GENDER=' ' ALL,
      INCOME*(N*F=8. MEAN*F=DOLLAR8. MEDIAN*F=DOLLAR8.) / RTS=10;
RUN;
```

The resulting table is shown in Output 24.24. As you can see, specifying the OUT= option has no effect on the table. The resulting data set is shown in Output 24.25. The three rows in our table have been turned into three observations: one for females, one for males, and one for ALL. Notice how these CLASS variables have retained their formats. The statistics we asked for have been turned into variables with names that indicate the analysis variable and the statistic. Long variable names come in handy here. Another nice feature is that the output data set retains additional decimal places for statistical results, ignoring the formats we specified. This can come in handy if you change your mind later about the precision you wish to display.

There are also three special variables. _TYPE_ identifies which CLASS variables are being reported in that row. _PAGE_ refers to the value of the page variable. _TABLE_ refers to the table number (both _PAGE_ and _TABLE_ are set to 1 because there is no page variable and there is only one table). For this simple table, with only one CLASS variable, the output data set is fairly straightforward. Where things can get a bit confusing is when you have more than one CLASS variable, as in the following example:

```
PROC TABULATE DATA=TEMP OUT=TABOUT ORDER=FORMATTED;
   CLASS GRADE GENDER;
   TABLE GENDER=' ', (GRADE ALL)*PCTN<GRADE ALL>*F=PCTPIC.
      / RTS=20 ROW=FLOAT;
RUN;
```

The resulting table is shown in Output 24.26. The resulting data set is shown in Output 24.27. As you can see, we no longer have one observation of data for each row in the table. What the OUT= option does is create a data set with an observation for each unique combination of the CLASS variables in your table. The layout of this data set is similar to the data sets produced by PROC MEANS or PROC SUMMARY.

In this case, that means one observation for each combination of GENDER and GRADE. There are also two observations for the total because ALL is a classification variable, too. The PctN_01 variable holds our percentages. The reason the PctN variable is numbered is that in tables with more complex percentages, the number helps you figure out which percentages belong to which denominator.

As you can see, understanding your output data set is a little tricky. And even this second example is fairly simple. Using the OUT= option with a complex table may create a data set that is too complex to meet your needs. Whenever you use this option, be sure to print and understand your output data set before you use it with other procedures.

Output 24.24

```
-----------------------------------------
|        |          Income            |
|        |----------------------------|
|        |    N   |  Mean  | Median  |
|--------+--------+--------+--------- |
|Female  |   2180 | $39,912| $34,003 |
|--------+--------+--------+--------- |
|Male    |   2129 | $23,597| $21,000 |
|--------+--------+--------+--------- |
|All     |   4309 | $31,852| $26,200 |
-----------------------------------------
```

Output 24.25

Obs	GENDER	_TYPE_	_PAGE_	_TABLE_	Income_N	Income_ Mean	Income_ Median
1	Female	1	1	1	2180	39912.50	34002.5
2	Male	1	1	1	2129	23597.45	21000.0
3		0	1	1	4309	31851.52	26200.0

Output 24.26

```
----------------------------------------------------------------------
|                  |                  Grade                  |       |
|                  |-----------------------------------------|       |
|                  |   A  |   B  |   C  |   D  |   F  |  All  |       |
|                  |------+------+------+------+------+-------|       |
|                  | PctN | PctN | PctN | PctN | PctN | PctN  |       |
|------------------+------+------+------+------+------+-------|       |
|Female            | 23.2%| 32.4%| 16.7%| 15.3%| 12.2%| 100.0%|       |
|------------------+------+------+------+------+------+-------|       |
|Male              | 24.1%| 33.7%| 14.5%| 17.1%| 10.3%| 100.0%|       |
----------------------------------------------------------------------
```

Output 24.27

Obs	GRADE	GENDER	_TYPE_	_PAGE_	_TABLE_	PctN_01
1	4	Female	11	1	1	23.241
2	3	Female	11	1	1	32.408
3	2	Female	11	1	1	16.785
4	1	Female	11	1	1	15.300
5	0	Female	11	1	1	12.266
6	4	Male	11	1	1	24.146
7	3	Male	11	1	1	33.793
8	2	Male	11	1	1	14.560
9	1	Male	11	1	1	17.136
10	0	Male	11	1	1	10.365
11	.	Female	01	1	1	100.000
12	.	Male	01	1	1	100.000

Output Delivery System

Probably the single biggest change in Version 7 is its new approach to SAS output. Sure, you can run all of your old Version 6 programs and produce the same old output in Version 7. But with the new Output Delivery System, you can take your old programs and use them to create colorful and visually appealing output. And you can get your output as HTML, Rich Text Format, PostScript, PCL, or Adobe PDF, in addition to plain text.

ODS captures your standard SAS output as it is produced and turns it into the format and style output you request. You can call ODS to process the output of your entire program, or just the output of one procedure. To show how ODS works, take a look at the following sample:

```
ODS HTML BODY='SAMPLE1.HTML';
PROC TABULATE DATA=TEMP;
   CLASS GENDER OCCUP;
   VAR INCOME;
   TABLE GENDER=' ', OCCUP*INCOME=' '*MEAN=' '*F=DOLLAR13.
      / BOX='Mean Income' RTS=10;
RUN;
ODS HTML CLOSE;
```

By adding two lines of code around a PROC TABULATE, you can turn a basic table into the HTML output shown in Output 24.28. Notice how the fonts have been changed to emphasize the row and column headings. And although you can't tell it from this output, the headings are blue.

And if you'd rather have output that you can use with a word processor, all you have to do is change the HTML option to RTF and provide a new filename, and the same code will produce a Rich Text Format file instead.[1] This output will be similar to the table shown in Output 24.28.

```
ODS RTF BODY='SAMPLE1.RTF';
(same PROC TABULATE code as previous example)
ODS RTF CLOSE;
```

The third type of output supported by ODS is Adobe Postscript.[1] This is a common format used to share documents across platforms and across the Internet. Again, all you have to change in your code is the ODS option. You will be prompted for the filename. Again, the output will be similar to the table shown in Output 24.28.

```
ODS PRINTER;
(same PROC TABULATE code as previous example)
ODS PRINTER CLOSE;
```

No matter what the format, the styles will be similar. The power of ODS is that you have complete control over every aspect of these styles. The following examples show how you can revise these styles or even build your own collection of styles to create customized output.

---

[1] RTF and PostScript output are experimental in Version 7. Expect limited functionality in Version 7, with fuller capabilities in the next release of SAS software.

Output 24.28

Mean Income	OCCUPATION			
	Managerial	Professional	Technical	Sales
FEMALE	$47,400	$38,626	$28,601	$38,299
MALE	$22,926	$25,774	$19,676	$18,845

Adding ODS Styles to Your Tables - I

The previous example showed what happens when you use the default ODS settings for PROC TABULATE output. But the power of ODS is that you are not limited to these settings. You can specify exactly how you want each part of your table to look.

For example, compare the table shown in Output 24.29 to the table shown in Output 24.30. The same PROC TABULATE code is used for both tables. In Output 24.29, the default ODS settings are used. In Output 24.30, the STYLE= (or S=) option is used throughout the PROC TABULATE code to modify the formatting of the resulting HTML table. Though you can't see it in this black and white output, the dark background areas in the heading and the "Overall" label are purple.

The code below shows how the table formatting is modified. The styles that were added to the code to produce the modified output are shown in light italics. The styles are added using the STYLE= (or S=) option, with the style attributes set off using {} or [] brackets. Style attributes consist of an attribute name, an equal sign operator, and the attribute setting or value. A list of the available attributes is in the ODS documentation.

```
ODS HTML BODY='SAMPLE2.HTML';
PROC TABULATE DATA=TEMP F=10.2
   S={FOREGROUND=BLACK BACKGROUND=WHITE FONT_WEIGHT=BOLD
      CELLWIDTH=80 JUST=C};
   CLASS JobClass Gender;
   CLASSLEV JobClass Gender / S={FOREGROUND=BLACK BACKGROUND=WHITE};
   CLASS Division;
   CLASSLEV Division / S={FOREGROUND=WHITE BACKGROUND=PURPLE};
   VAR JobSatisfaction;
   TABLE Gender=' '*JobClass=' '
      ALL={LABEL='Overall'
            S={FOREGROUND=WHITE BACKGROUND=PURPLE JUST=R}},
      MEAN={LABEL='Mean Job Satisfaction Rating'
            S={FOREGROUND=WHITE BACKGROUND=PURPLE}}*
      Division=' '*JobSatisfaction=' '
      / BOX={LABEL='Cuppa Coffee Co.'
             S={FOREGROUND=WHITE BACKGROUND=PURPLE VJUST=T JUST=L}}
             S={RULES=None CELLSPACING=0 CELLPADDING=10};
   RUN;
ODS HTML CLOSE;
```

You may have noticed a new statement used in the code: the CLASSLEV statement. This is like a CLASS statement, except that it is used to attach styles to the classification variable *values*. Style attributes used include the FOREGROUND and BACKGROUND colors and the horizontal (JUST) and vertical (VJUST) justification settings. The RULES, CELLSPACING, and CELLPADDING settings were used to get rid of the table dividers.

While a detailed explanation of each of these settings is beyond the scope of this book, you should be able to get a basic idea of how the settings affect the table by comparing Output 24.29 and Output 24.30 and looking at the code.

Output 24.29

Cuppa Coffee Co.		Mean Job Satisfaction Rating			
		Northeast	Southeast	Midwest	West
Female	Managerial	3.45	4.00	4.00	3.17
	Clerical	3.02	4.96	4.04	3.00
	Manufacturing	4.03	4.24	3.10	2.16
Male	Managerial	4.07	3.00	5.00	4.00
	Clerical	4.00	3.00	5.00	4.00
	Manufacturing	3.19	4.31	4.06	3.06
Overall		3.45	4.21	3.61	2.74

Output 24.30

Cuppa Coffee Co.		Mean Job Satisfaction Rating			
		Northeast	Southeast	Midwest	West
Female	Managerial	3.45	4.00	4.00	3.17
	Clerical	3.02	4.96	4.04	3.00
	Manufacturing	4.03	4.24	3.10	2.16
Male	Managerial	4.07	3.00	5.00	4.00
	Clerical	4.00	3.00	5.00	4.00
	Manufacturing	3.19	4.31	4.06	3.06
	Overall	3.45	4.21	3.61	2.74

Adding ODS Styles to Your Tables - II

The following example shows off a number of style tricks. Because it is fairly complex, the code and examples are spread out over four pages. To see the output, flip forward and look at Output 24.31 and Output 24.32. Don't worry about understanding every detail of this example. It's intended to whet your appetite for the possibilities of PROC TABULATE and ODS.

The first step in producing these tables is to set up several user-defined formats. ODS lets you put formats to use to help control complex table designs. There are three formats used in the table:

```
ODS HTML BODY='SAMPLE3.HTML';
PROC FORMAT;
    VALUE DIVFLY  1='NY,NJ,MD,CN,MA,RI'
                  2='VA,NC,SC,FL,GA,AL'
                  3='IA,NE,KS,OH,IN'
                  4='WA,OR,CA,ID,NV,UT';
    VALUE GENDGIF 1='female.gif'
                  2='male.gif';
    VALUE WATCHIT 0-<3='RED'
                  3-<4='YELLOW'
                  4-high='DARKGREEN';
    RUN;
```

The DIVFLY format supplies text values that describe the states included in each division. If you look at Output 24.32, you can see that these values are used to produce dynamic labels for the columns. As the cursor passes over one of the column headings, a popup appears with a definition. This effect is achieved using the DIVFLY format and the FLYOVER attribute, which are explained below.

The GENDGIF format is used to indicate two graphics files. For GENDER=1, this format points to an image file FEMALE.GIF. For GENDER=2, this format indicates that the file MALE.GIF should be used. You can see these images in Output 24.31. These images are applied to the table using the GENDGIF format and the POSTIMAGE attribute, which are explained below.

The third format, WATCHIT, is used to specify different text colors for three ranges of table cell values. This allows low values to be shown with red text, medium values to be shown with yellow text, and high values to be shown with green text (though it is hard to see this in the black and white table shown in Output 24.31). The colors are applied to the table using the WATCHIT format and the FOREGROUND attribute, which are explained below.

The next piece of the code is the PROC TABULATE statement. This statement, like the rest of this example, uses a number of style attributes. The styles specified in this statement affect the table cells.

```
PROC TABULATE DATA=TEMP F=10.2
    S={FOREGROUND=WATCHIT. BACKGROUND=LIGHTBLUE
FONT_SIZE=4 FONT_WEIGHT=BOLD CELLWIDTH=80 JUST=C};
```

In this case, the FOREGROUND is set to the WATCHIT format, which means that the cells will have red, yellow, and green values, depending on the value displayed in the cell. The BACKGROUND attribute controls the color of the cell background. The FONT_SIZE and FONT_WEIGHT control the font used to display the cell values. The CELLWIDTH setting is used to create a uniform cell width. Without this setting, columns with narrower headings would have narrower columns. This setting forces all columns to have the same width. Finally, the JUST setting calls for the values to be centered in each cell.

The next part of the code controls the row and column headings. The CLASS statement applies styles to the CLASS variable headings (in this case, no styles are applied), and the CLASSLEV statements apply styles to the CLASS variable *value* headings.

```
CLASS JobClass Gender Division;
CLASSLEV JobClass  / S={FOREGROUND=WHITE BACKGROUND=BLUE};
CLASSLEV Gender   / S={FOREGROUND=WHITE BACKGROUND=BLUE VJUST=T
                      POSTIMAGE=GENDGIF. CELLWIDTH=80};
CLASSLEV Division  / S={FOREGROUND=WHITE BACKGROUND=DARKBLUE
                      FLYOVER=DIVFLY.};
```

In this example, no styles are applied in the CLASS statement. This is because the label for each CLASS variable is going to be blanked out in the TABLE statement below, so there is no overall heading for JobClass, Gender, or Division. Instead, the CLASS variable values are used for row and column headings. The CLASSLEV statements are used to apply formats to these headings.

The first CLASSLEV statement is very simple. It sets the text color for the JobClass values to white and the background for these values to blue.

The next CLASSLEV statement is more complex. Again the colors are set to white text on blue. In addition, the width of the heading is set to 80, and the vertical justification (VJUST) is set to top. This means that the text is displayed starting at the top of the heading. The most interesting part of this CLASSLEV statement is the POSTIMAGE attribute. This refers to the GENDGIF format that is explained above. This attribute tells ODS and PROC TABULATE to display the image indicated by the format immediately following the CLASS variable value in each heading. This creates the images of a man and a woman that are shown in the row headings in Output 24.31.

The final CLASSLEV statement again sets the colors to white on blue. Then it uses the FLYOVER attribute to indicate that the resulting table is dynamic. Whenever the mouse cursor is passed over one of the Division values, the format DIVFLY is used to display a definition of the division. You can see how this works in Output 24.32. One warning: This and other advanced features may not work in all browsers.

The next portion of code includes a standard VAR statement and then a complex TABLE statement. Numerous style attributes are used to override the formats previously set up in the PROC TABULATE and CLASSLEV statements. This can be a little confusing. Basically, all you have to remember is that there is a hierarchy of styles. Styles specified in the PROC TABULATE, CLASS, CLASSLEV, and VAR statements are overridden by styles specified in the TABLE statement.

```
VAR JobSatisfaction;
TABLE Gender=' '*JobClass=' '
   ALL={LABEL='Overall'
        S={FOREGROUND=WHITE BACKGROUND=BLUE JUST=R}},
   Mean={LABEL='Mean Job Satisfaction Rating'
        S={FOREGROUND=WHITE BACKGROUND=DARKBLUE}}*
   Division=' '*JobSatisfaction=' '
      / BOX={LABEL='Cuppa Coffee Co.' S={FOREGROUND=WHITE
              BACKGROUND=DARKBLUE JUST=L VJUST=T
              PREIMAGE='LOGO.GIF' CELLWIDTH=144}}
              S={RULES=None CELLSPACING=0 CELLPADDING=10};
RUN;
ODS HTML CLOSE;
```

The styles for the ALL keyword and the MEAN statistic are used to create the white-on-blue color scheme used in the row and column headings. The styles used with the BOX option are used to extend this color scheme to the table BOX. The PREIMAGE attribute is used to display the company logo in the BOX. The JUST, VJUST, and CELLWIDTH attributes are used to orient the logo in a properly sized BOX.

The last S= is used to apply overall styles to the table grid. This style setting is applied as a TABLE statement option. In this case, the table does have any grid lines or separation between the cells (RULES=None and CELLSPACING=0), and there are 10 units of padding between the table values and labels and the edges of each table cell (CELLPADDING=10).

After all of this code, you end up with the tables shown in Output 24.31 and Output 24.32. This is probably a lot more work than you would want to go through for most tables. But if you have an important table that you want to dress up, experiment with the techniques shown in this example.

Output 24.31

Cuppa Coffee Co.	Mean Job Satisfaction Rating			
	Northeast	Southeast	Midwest	West
Female Managerial	3.45	4.00	4.00	3.17
Clerical	3.02	4.96	4.04	3.00
Manufacturing	4.03	4.24	3.10	2.16
Male Managerial	4.07	3.00	5.00	4.00
Clerical	4.00	3.00	5.00	4.00
Manufacturing	3.19	4.31	4.06	3.06
Overall	3.45	4.21	3.61	2.74

Output 24.32

Cuppa Coffee Co.	Mean Job Satisfaction Rating			
	Northeast	Southeast	Midwest IA,NE,KS,OH,IN	West
Female Managerial	3.45	4.00	4.00	3.17
Clerical	3.02	4.96	4.04	3.00
Manufacturing	4.03	4.24	3.10	2.16
Male Managerial	4.07	3.00	5.00	4.00
Clerical	4.00	3.00	5.00	4.00
Manufacturing	3.19	4.31	4.06	3.06
Overall	3.45	4.21	3.61	2.74

ODS Styles for Tables

As you have probably noticed in the previous examples, there are many style attributes that can be used with PROC TABULATE, and there are many places that you can use them. In an attempt to sort this out, Table 24.1 lists each of the places where you can apply a style and the part of the table affected. For a list of the style attributes and their meanings, consult the ODS documentation.

Table 24.1

Syntax	*Part of the table that is affected*
`PROC TABULATE S={...};`	data cells

`CLASS varname / S={...};`
or, in dimension expression,
`varname={LABEL='string' S={...}}`
	heading (page heading, row heading or column heading) for the named CLASS variable

`CLASSLEV varname / S={...};`
or, in dimension expression,
`varname={LABEL='string'`
`S(CLASSLEV)={...}}`
	classification values for the named CLASS variable

`VAR varname / S={...};`
or, in dimension expression,
`varname={LABEL='string' S={...}}`
	heading (page heading, row heading, or column heading) for the named VAR variable

`KEYWORD ALL|statistic / S={...};`
or, in dimension expression,
`ALL|statistic={LABEL='string' S={...}}`
	heading (page heading, row heading, or column heading) for ALL or the named statistic

`TABLE <page dimension,> <row dimension,>` `column dimension / S={};`	table border, table rules, cell spacing
`BOX={LABEL='string' S={}}`	table BOX

The explanation above is greatly simplified. There are aspects of a table's formatting that can be specified in more that one location. In general, settings applied to specific variables or statistics in the TABLE statement override settings applied outside the TABLE statement. Another thing to keep in mind is that not all ODS style attributes can be applied to PROC TABULATE, and there are limits on which attributes can be applied where.

Attributes that can be applied to the TABLE statement include: BackGround, BackGroundImage, BorderColor, BorderColorDark, BorderColorLight, BorderWidth, CellPadding, CellSpacing, Font_Face, Font_Size, Font_Style, Font_Weight, Font_Width, ForeGround, Frame, Just, OutputWidth, PostHTML, PostImage, PostText, PreHTML, PreImage, PreText, ProtectSpecialChars, and Rules.

Attributes that can be applied everywhere else include: AsIs, BackGround, BackGroundImage, BorderColor, BorderColorDark, BorderColorLight, BorderWidth, CellHeight, CellPadding, CellSpacing, CellWidth, Flyover, Font_Face, Font_Size, Font_Style, Font_Weight, Font_Width, ForeGround, Just, NoBreakSpace, OutputWidth, PostHTML, PostImage, PostText, PreHTML, PreImage, PreText, ProtectSpecialChars, VisitedLinkColor, and Vjust.

The interaction of these style settings is very complex, and building a table with complex formatting may involve some trial and error experimenting on your part.

ODS Style Templates for Tables

While most PROC TABULATE styles are modified within the PROC TABULATE code, this is not true of all procedures in Version 7. For most procedures, these styles are controlled by style templates.

To build a style template, you use a new SAS procedure released in Version 7. PROC TEMPLATE allows you to specify a standard format for each part of each procedure's output. The following example shows how you can use PROC TEMPLATE to set up a new PROC TABULATE style template.

Building a style template is a great way to create a standard look and feel for all of the tables that you produce. If you would like all of your tables to use the same color scheme and show the company logo, you can create a custom style template to use with PROC TABULATE instead of the default style template.

The following code builds the style template used in Output 24.33. The PROC TABULATE code is the same as was used in the original ODS example shown in Output 24.28. The changes to the style template include inserting a default logo and page title, changing the table color scheme to dark and light shades of coffee brown, and removing the table cell borders.

The first thing you have to do is give your style template a name. Then you specify that it be based on the default style template. This means that you can avoid specifying everything except the things that you want to change.

```
PROC TEMPLATE;
   DEFINE STYLE STYLES.MYSTYLE; PARENT=STYLES.DEFAULT;
```

The next section sets up the standard colors that will be used throughout the table. The colors are represented by hexadecimal codes. This numbering scheme is commonly used in HTML code. If you want to do a lot of ODS manipulation of your output, you may find it useful to review an introductory text or tutorial on HTML.

```
STYLE COLORS /
   'HEADERS'=CX663300
   'CELLS'=CXCC9966
   'TEXTBLK'=CX000000
   'TEXTWHT'=CXFFFFFF;
```

The next line sets up the header that will be used for all output page headings. In this case, the company logo and name will be displayed at the top of every page. This is achieved by writing a snippet of HTML code to create the heading, and then telling SAS to put this at the top of each page by using the PREHTML attribute.

```
STYLE BODY /
   PREHTML='<H1><IMG SRC="C:\TEMP\LOGO.GIF">Cuppa Coffee Co.</H1>';
```

The next section of code is used to specify the styles to use for the row and column heaers. These styles could be applied in the PROC TABULATE code. The advantage of doing it in the PROC TEMPLATE is that you only have to do it once, and then all of your tables will have the same look.

```
STYLE HEADER FROM HEADERSANDFOOTERS /
   BACKGROUND=COLORS('HEADERS')
   FOREGROUND=COLORS('TEXTWHT')
   VJUST=T;
STYLE ROWHEADER FROM HEADERSANDFOOTERS /
   BACKGROUND=COLORS('HEADERS')
   FOREGROUND=COLORS('TEXTWHT');
```

The last section of code is used to specify some general styles that will be used for the entire table. The combination of CELLSPACING=0, BORDERWIDTH=0, and RULES=NONE is used to remove all grid lines from the table. Because we have already specified styles for the row and column headers, the BACKGROUND and FOREGROUND colors specified here are applied to the data cells only.

```
STYLE TABLE FROM TABLE /
    BACKGROUND=COLORS('CELLS')
    FOREGROUND=COLORS('TEXTBLK')
    CELLSPACING=0
    BORDERWIDTH=0
    RULES=NONE;
END; RUN;
```

In this example, the Header, RowHeader, and Table styles are modified. The other styles that affect PROC TABULATE are the Data and Caption styles. The Data style affects how the values in the table cells are displayed. The Caption style controls how the page dimension labels (if you have a three-dimensional table) are displayed.

To apply the above style template to your PROC TABULATE table, specify it in the ODS statement, as in the following code:

```
ODS HTML BODY='C:\TEMP\MYTABLE.HTML' STYLE=MYSTYLE;
PROC TABULATE DATA=TEMP;
    CLASS GENDER OCCUP;
    VAR INCOME;
    TABLE GENDER=' ', OCCUP*INCOME=' '*MEAN=' '*F=DOLLAR13.
        / BOX='Mean Income' RTS=10;
RUN;
ODS HTML CLOSE;
```

The resulting table is shown in Output 24.33. But building just a single table was not the point of this exercise. Now that you have defined a style template, all you have to do to apply it to a new table is change the PROC TABULATE code that you include between the two ODS statements. Output 24.34 shows the same style template applied to a different table. This style template also works with output from other procedures (PROC PRINT, PROC FREQ, and so on).

Output 24.33

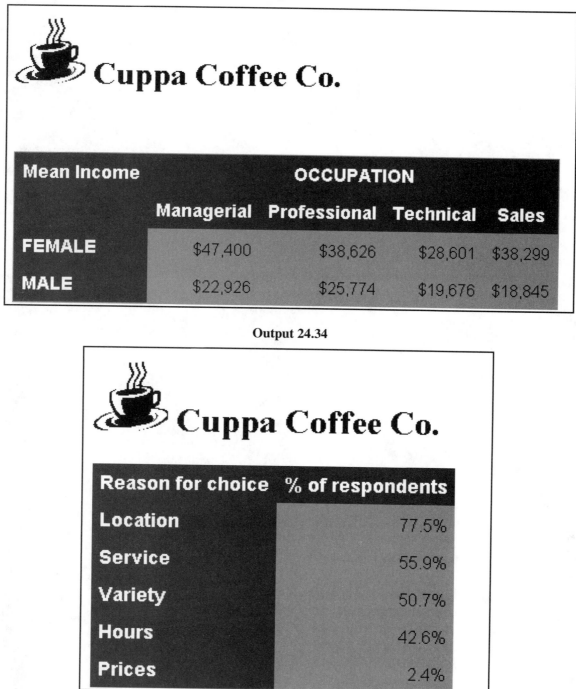

Output 24.34

By using a single STYLE statement, you can create a customized table but without having to clutter up your PROC TABULATE code with all of those STYLE= settings. If you find yourself making the same customizations over and over again, you should invest the time to set up a custom style template using PROC TEMPLATE.

References

This book is designed to guide you step by step through the table production process, but there is a lot of information on PROC TABULATE that is beyond the scope of this book to present.

The following page lists a number of useful references: SAS manuals related to PROC TABULATE, background reading for newer SAS users, and additional tips and techniques from other SAS users.

SAS Manuals

SAS Institute Inc. (1990), *SAS Guide to TABULATE Processing, Second Edition,* Cary, NC: SAS Institute Inc.

SAS Institute Inc. (1990), *SAS Language: Reference, Version 6, First Edition,* Cary, NC: SAS Institute Inc.

SAS Institute Inc. (1990), *SAS Procedures Guide, Version 6, Third Edition,* Cary, NC: SAS Institute Inc.

SAS Institute Inc. (1996), *Getting Started with the SAS System Using SAS/ASSIST Software, Version 6, Second Edition,* Cary, NC: SAS Institute Inc.

SAS Institute Inc. (1997), *SAS Macro Language: Reference, First Edition,* Cary, NC: SAS Institute Inc.

Background Reading for Beginners

Delwiche, L. and Slaughter, S. (1995), *The Little SAS Book: A Primer,* Cary, NC: SAS Institute Inc.

Pass, R. (1997), "PROC REPORT: An Introduction to the Batch Language," *Proceedings of the Twenty-Second Annual SAS Users Group International Conference,* 22, 682-691.

More Tips and Techniques for Intermediate and Advanced Users

Bruns, D. (1997), "The Truly Advanced Features of the TABULATE Procedure," *Proceedings of the Twenty-Second Annual SAS Users Group International Conference,* 22, 251-256.

Gravely, A. (1998), *Your Guide to Survey Research Using the SAS System,* Cary, NC: SAS Institute Inc.

Li, C. and Sun, J. (1998), "Using Hyperlink to Organize SAS HTML Output," *Proceedings of the Twenty-Third Annual SAS Users Group International Conference,* 23, 986-990.

Mason, P. (1996), *In the Know ... Tips and Techniques from Around the Globe,* Cary, NC: SAS Institute Inc.

Noga, S. and Abolafia, J. (1998), "The Tabulate Procedure: One Step Beyond the Final Chapter," *Proceedings of the Twenty-Third Annual SAS Users Group International Conference,* 23, 839-844.

Noga, S. and Zhao, D. (1997), "Odds Ratios in a Tabular Presentation," *Proceedings of the Twenty-Second Annual SAS Users Group International Conference,* 22, 970-975.

SAS Institute Inc. (1994), *Reporting From the Field: SAS Software Experts Present Real-World Report-Writing Applications,* Cary, NC: SAS Institute Inc.

Tien, P., Lin, T., and McGranachan, M. (1997), "Some Tips and Examples for Using PROC TABULATE," *Proceedings of the Twenty-Second Annual SAS Users Group International Conference,* 22, 1024-1029.

Index

Call your local SAS® office to order these other books and tapes available through the Books by Users℠ program:

An Array of Challenges — Test Your SAS® Skills
by **Robert Virgile**..Order No. A55625

Applied Multivariate Statistics with SAS® Software,
Second Edition
by **Ravindra Khattree**
and **Dayanand N. Naik**..............................Order No. A56903

Applied Statistics and the SAS® Programming Language,
Fourth Edition
by **Ronald P. Cody**
and **Jeffrey K. Smith**................................Order No. A55984

Beyond the Obvious with SAS® Screen Control Language
by **Don Stanley** ....Order No. A55073

Carpenter's Complete Guide to the SAS® Macro Language
by **Art Carpenter**......................................Order No. A56100

The Cartoon Guide to Statistics
by **Larry Gonick**
and **Woollcott Smith**................................Order No. A55153

Categorical Data Analysis Using the SAS® System
by **Maura E. Stokes, Charles S. Davis,**
and **Gary G. Koch**....................................Order No. A55320

Common Statistical Methods for Clinical Research with
SAS® Examples
by **Glenn A. Walker**..................................Order No. A55991

Concepts and Case Studies in Data Management
by **William S. Calvert**
and **J. Meimei Ma**....................................Order No. A55220

Efficiency: Improving the Performance of Your SAS®
Applications
by **Robert Virgile**......................................Order No. A55960

Essential Client/Server Survival Guide, Second Edition
by **Robert Orfali, Dan Harkey,**
and **Jeri Edwards**....................................Order No. A56285

Extending SAS® Survival Analysis Techniques for
Medical Research
by **Alan Cantor**..Order No. A55504

A Handbook of Statistical Analyses Using SAS®
by **B.S. Everitt**
and **G. Der** ....Order No. A56378

The How-To Book for SAS/GRAPH® Software
by **Thomas Miron** ....Order No. A55203

In the Know ... SAS® Tips and Techniques From
Around the Globe
by **Phil Mason** ....Order No. A55513

Integrating Results through Meta-Analytic Review Using
SAS® Software
by **Morgan C. Wang** and
Brad J. Bushman ....Order No. A55810

Learning SAS® in the Computer Lab
by **Rebecca J. Elliott** ....Order No. A55273

The Little SAS® Book: A Primer
by **Lora D. Delwiche** and
Susan J. Slaughter ....Order No. A55200

The Little SAS® Book: A Primer, Second Edition
by **Lora D. Delwiche** and
Susan J. Slaughter ....Order No. A56649
(updated to include Version 7 features)

Logistic Regression Using the SAS System:
Theory and Application
by **Paul D. Allison** ....Order No. A55770

Mastering the SAS® System, Second Edition
by **Jay A. Jaffe** ....Order No. A55123

Multiple Comparisons and Multiple Tests Using
the SAS® System
by **Peter H. Westfall, Randall D. Tobias,**
Dror Rom, Russell D. Wolfinger,
and **Yosef Hochberg** ....Order No. A56648

The Next Step: Integrating the Software Life Cycle with
SAS® Programming
by **Paul Gill** ....Order No. A55697

Painless Windows 3.1: A Beginner's Handbook for
SAS® Users
by **Jodie Gilmore** ....Order No. A55505

Painless Windows: A Handbook for SAS® Users
by **Jodie Gilmore** ....Order No. A55769
(for Windows NT and Windows 95)

Painless Windows: A Handbook for SAS® Users,
Second Edition
by **Jodie Gilmore** ....Order No. A56647
(updated to include Version 7 features)

PROC TABULATE by Example
by **Lauren E. Haworth** ....Order No. A56514

Professional SAS® Programmers Pocket Reference,
Second Edition
by **Rick Aster** ....Order No. A56646

Professional SAS® Programming Secrets, Second Edition
by **Rick Aster**
and **Rhena Seidman** ....Order No. A56279

Professional SAS® User Interfaces
by **Rick Aster** ....Order No. A56197

*Welcome * Bienvenue * Willkommen * Yohkoso * Bienvenido*

SAS® Publications Is Easy to Reach

Visit our SAS Publications Web page located at www.sas.com/pubs

You will find product and service details, including

- **sample chapters**
- **tables of contents**
- **author biographies**
- **book reviews**

Learn about

- **regional user groups conferences**
- **trade show sites and dates**
- **authoring opportunities**
- **custom textbooks**

Order books with ease at our secured Web page!

Explore all the services that Publications has to offer!

Your Listserv Subscription Brings the News to You Automatically

Do you want to be among the first to learn about the latest books and services available from SAS Publications? Subscribe to our listserv **newdocnews-l** and automatically receive the following once each month: a description of the new titles, the applicable environments or operating systems, and the applicable SAS release(s). To subscribe:

1. Send an e-mail message to **listserv@vm.sas.com**

2. Leave the "Subject" line blank

3. Use the following text for your message:

　　　subscribe newdocnews-l *your-first-name your-last-name*

　　For example: subscribe newdocnews-l John Doe

　　Please note: newdocnews-l ◄——— that's the letter "l" not the number "1".

For customers outside the U.S., contact your local SAS office for listserv information.

Create Customized Textbooks Quickly, Easily, and Affordably

SelecText® offers instructors at U.S. colleges and universities a way to create custom textbooks for courses that teach students how to use SAS software.

For more information, see our Web page at **www.sas.com/selectext**, or contact our SelecText coordinators by sending e-mail to **selectext@sas.com**.

You're Invited to Publish with SAS Institute's User Publishing Program

If you enjoy writing about SAS software and how to use it, the User Publishing Program at SAS Institute Inc. offers a variety of publishing options. We are actively recruiting authors to publish books, articles, and sample code. Do you find the idea of writing a book or an article by yourself a little intimidating? Consider writing with a co-author. Keep in mind that you will receive complete editorial and publishing support, access to our users, technical advice and assistance, and competitive royalties. Please contact us for an author packet. E-mail us at **sasbbu@sas.com** or call 919-677-8000, then press 1-6479. See the SAS Publications Web page at **www.sas.com/pubs** for complete information.

Read All about It in *Authorline*®!

Our User Publishing newsletter, *Authorline*, features author interviews, conference news, and informational updates and highlights from our User Publishing Program. Published quarterly, *Authorline* is available free of charge. To subscribe, send e-mail to **sasbbu@sas.com** or call 919-677-8000, then press 1-6479.

See *Observations*®, Our Online Technical Journal

Feature articles from *Observations*®: *The Technical Journal for SAS*® *Software Users* are now available online at **www.sas.com/obs**. Take a look at what your fellow SAS software users and SAS Institute experts have to tell you. You may decide that you, too, have information to share. If you are interested in writing for *Observations*, send e-mail to **sasbbu@sas.com** or call 919-677-8000, then press 1-6479.

Book Discount Offered at SAS Public Training Courses!

When you attend one of our SAS Public Training Courses at any of our regional Training Centers in the U.S., you will receive a 15% discount on any book orders placed during the course. Each course has a list of recommended books to choose from, and the books are displayed for you to see. Take advantage of this offer at the next course you attend!

SAS Institute Inc.
SAS Campus Drive
Cary, NC 27513-2414
Fax 919-677-4444

E-mail: sasbook@sas.com
Web page: www.sas.com/pubs
To order books, call Fulfillment Services at 800-727-3228*
For other SAS Institute business, call 919-677-8000*

* **Note:** Customers outside the U.S. should contact their local SAS office.

SAS®